SMALL TECH

ELECTRONIC MEDIATIONS

KATHERINE HAYLES, MARK POSTER, AND SAMUEL WEBER
SERIES EDITORS

SMALL TECH

The Culture of Digital Tools

BYRON HAWK, DAVID M. RIEDER,
AND OLLIE OVIEDO, EDITORS

ELECTRONIC MEDIATIONS, VOLUME 22

UNIVERSITY OF MINNESOTA PRESS
MINNEAPOLIS • LONDON

Chapter 1 originally appeared as *Artphoto* (Buharest), 2003. http://www.artphoto.ro/

A version of chapter 12 was previously published as Steve Mann, Jason Nolan, and Barry Wellman, "Sousveillance: Holding a Technological Mirror up to Society," *Surveillance and Society* 1 (June 2003). http://www.surveillance-and-society.org/articles1(3)/sousveillance.pdf.

Material from chapter 13 appeared as Jim Bizzochi, "Video as Ambience: Reception and Aesthetics of Flat-Screen Video Display," *Crossings: Electronic Journal of Art and Technology* 4, no. 1 (December 04). http://crossings.tcd.ie/issues/4.1/Bizzocchi/.

Published by the University of Minnesota Press
111 Third Avenue South, Suite 290
Minneapolis, MN 55401-2520
http://www.upress.umn.edu

Library of Congress Cataloging-in-Publication Data

Small tech : the culture of digital tools / Byron Hawk, David M. Rieder, and Ollie Oviedo, editors.
    p. cm. — (Electronic mediations ; v. 22)
    Includes bibliographical references.
    ISBN: 978-0-8166-4977-8 (hc : alk. paper)
    ISBN-10: 0-8166-4977-4 (hc : alk. paper)
    ISBN: 978-0-8166-4978-5 (pb : alk. paper)
    ISBN-10: 0-8166-4978-2 (pb : alk. paper)
    1. Telematics. 2. Miniature electronic equipment—Social aspects. 3. Pocket computers—Social aspects. 4. Household electronics—Social aspects. 5. Cyberspace. 6. Mass media—Technological innovations. 7. Digital music players—Social aspects. 8. Cellular telephones—Social aspects. I. Hawk, Byron. II. Rieder, David M. III. Oviedo, Ollie O.
    TK5105.6.S62 2008
    303.48'3—dc22 2007033558

Contents

Acknowledgments

As anyone who has edited a collection of other people's work knows, the process is often a long one. We would first like to thank all of the authors who stuck with the project, endured our multiple requests for revision, and contributed to the emergence of what the collection has become. We especially thank James Inman, who contributed early in the process; series editors N. Katherine Hayles, Mark Poster, and Samuel Weber for seeing value in the project; Douglas Armato and Katie Houlihan at the University of Minnesota Press for insight and advice; and all of the reviewers for valuable feedback.

Introduction

On Small Tech and Complex Ecologies *interrelated*

Byron Hawk and David M. Rieder

———————

(2000s)

In the past ten years, scholars across various disciplines have been thinking about the impact of new media on various social and aesthetic sites of engagement. In his introduction to *The New Media Reader*, "New Media from Borges to HTML," Lev Manovich makes the distinction between two paths of research that he labels cyberculture and new media. Writers investigating cyberculture examine various social phenomena related to the *cyber-culture* Internet and other forms of networked communication. Issues involving online community, online gaming, identity, gender, ethnicity, and online ethnography are common in this field. New media researchers, on the other hand, focus on cultural objects influenced by *new media* contemporary computing that include nonnetworked cultural objects. New media, in this context, comes to mean a variety of aesthetic and artistic productions such as installation art, computer games, and video- and image-based works, as well as multimedia projects and virtual reality. In short, the one emphasizes the social aspects of virtual space, while the other emphasizes the aesthetic possibilities of the virtual and their relation to physical installations. Manovich charts the divergent development of these two modes of inquiry *different regional focuses* in terms of territory. Cyberculture studies have been prominent in the United States because the Internet as a social phenomenon arose more quickly there than in other countries. New media studies emerged in Europe not only because the Internet was slow to come and lacked the strong corporate backing seen in the United States but also because public funding for the arts was much more prevalent, which created the conditions for developing larger installation artworks not necessarily tied to the Internet.

With the emergence of "small tech"—iPods, cell phones, digital cameras, personal digital assistants (PDAs)—digital culture is no longer confined to these two domains. The next wave of new media studies will need to examine the ecological interrelationships among the virtual space of the Internet, the enclosed space of the installation, and the open space of everyday life. Small tech usually derives its meaning from science and engineering research into areas such as microelectromechanics (MEMs) and nanotechnology, which conjures up images of fields as diverse as telecommunications, energy, and biotechnology. But in more vernacular contexts, small tech signals the emerging interconnected social, cultural, aesthetic, and educational contexts that will be brought into new relations through the development of smaller hardware and the recontextualization of preexisting software. Small tech not only signals a shift in perspective from industrial to postindustrial technology, as the Internet did, but also highlights the changing nature of our material relations to technologies:

> Small tech is expected to have a greater impact on our lives in the next 10 years than the Internet has had in the last 10 years. As technology allows us to manipulate smaller components, the potential for new and improved products will have no boundaries. A computer the size of a cell phone, a cell phone the size of a hearing aid, and a hearing aid the size of a pinhead. ("About Small Tech")

Small tech, then, runs parallel with a general process of miniaturization. Increasingly miniaturized components and goods redefine the threshold between materiality and the conceptual, cognitive, and affective dimensions of everyday life. In order to investigate these emerging relationships, *Small Tech: The Culture of Digital Tools* brings together scholars from a variety of fields and countries to assess the implications of a variety of specific technologies and the diverse ecologies that they engage and make possible.

Small Tech and Weightless Economics

Even though Marshall McLuhan's trope of the global village persists in both popular and scholarly depictions as the end goal of technological progress, the emergence of small tech has problematized the implied values of this unified model. In McLuhan's descriptions of an emerging global village, media technologies will bring diverse peoples together as one. If everyone is participating in the same media culture, then there is the potential for equality. As Steven Shaviro puts it in the opening of *Connected*, "All its pieces fit seamlessly together, and each problem receives an optimal solution" (4). During the height of the Internet bubble in the 1990s, proponents of economic globalism adopted this wholistic logic to promote the vision of a one-world New Economy in which technology would surpass the noise of the local and bring all people into instant communication. Kevin Kelly, George Gilder, and even Bill Gates developed this McLuhanesque ideal, using its progressive implications to promote a zeal for technological innovation. The connections to capitalist values, however, are almost too convenient. A global ecology implies both the natural harmony of a survival of the economically fittest as well as the displacement of that fitness onto the system rather than the individual or corporation. But, as economist Diane Coyle notes, our contemporary situation has not evolved in quite the way the proponents of globalism predicted. As she argues in *The Weightless World*, the logic of the global village is still understood on an industrial scale, but the economy is turning toward a weightless world.

A more productive vantage point from which to examine the contemporary context created by emerging small tech is to associate it with the relatively new field of weightless economics, which is concerned with the economic value of immaterial goods, services, and techniques that increasingly define the leading edges of first-world economies. Cultural theorist and commentator Jeremy Rifkin writes, "If the industrial era was characterized by the amassing of physical capital and property, the new era prizes intangible forms of power bound up in bundles of information and intellectual assets" (30). The tallest buildings, the largest ships, and the heaviest, most physically powerful machinery are no longer the centerpieces of American economic value and profit. Up through the end of World War II, size and weight were indicators of strength and security: the amassing of physical property was a sign of a global power. Today, over a half-century after World War II, this industrial value has been replaced by a reverse trend toward weightlessness and miniaturization. Since the end of World War II, nations like the United States have been spending increasing amounts of money on services, experiences, and a wide range of immaterial products and less money on physical, tangible goods.

According to Coyle, the credit for first characterizing the American economy as increasingly weightless goes to Alan Greenspan. Greenspan first introduced the concept of weightlessness in 1996, noting that, "While the weight of current economic output is probably only modestly higher than it was half a century ago, value added adjusted for price change has risen well over three-fold" ("Remarks"). In other words, the United States is three times more valuable now than it was just after World War II, but it still weighs the same. For an economist, this is something to ponder. As Coyle claims, "The concept of the balance of trade started out as a literal description of how the quantity of imports and exports was assessed: the weight of imports on one side of the scales and the weight of exports on the other" (1). Even as late as 1985, according to Coyle, this concept was still in use in the United Kingdom. A decade later, however, digital technologies made the measure almost useless. Computers had become so powerful that their weight was no indication of their processing capacity or value. "A single imported greeting card with a microchip that plays Happy Birthday when the card is opened," for example, "contains more computer power than existed on the planet 50 years ago. It weighs a gram or so" (Coyle 1). This shift in weight creates radically different material relations among nations, increasing the complexity but not necessarily the equality of their relationships. Consequently, Greenspan challenges economists to focus more explicitly on the far-reaching implications of weightlessness.

This same call should also be made to humanities scholars. The humanities have generally followed McLuhan in seeing technologies as extensions of human capacities. But as Manovich notes in *The Language of New Media*, scale becomes an essential component of digital technologies, which allows them to affect relations on a scale as large as nations and as small as ones and zeros (38–40). Miniaturization, then, represents a de-centering of McLuhan's human-centered model. The issue becomes not the gun as the extension of the human body, as Richard Doyle points out, but the mouse as the extension of the hand into the world of ones and zeros (41). Rather than extending human power, the hand "jacks in" to the power of the network itself. Small tech—both the weightless ones and zeros of software and the increasing miniaturization of hardware—extends power to include the fragment, the individual, the corporation, and the network. Power isn't simply displaced onto the system but is distributed across the system at all levels of scale. The power of the network not only allows corporations to capitalize on it but also allows anyone connected to it to utilize and ultimately affect the network. This doesn't mean, however, that power relations are equal as in the early depictions of the global village. Power in the era of small tech is increasingly available, but it is also increasingly uneven.

Small tech highlights the complexity of the threshold between the material world of big, physical things and the virtual worlds of conceptual, affective communication and calls for the attention of humanities scholarship. Cultural theorists such as Gilles Deleuze,

[handwritten margin note: miniaturization of tech + digital culture like an omni – microscope/ telescope, opening up scale at both ends]

Félix Guattari, Brian Massumi, Alphonso Lingis, Pierre Lévy, and N. Katherine Hayles are looking at the ways in which the weightless world of virtuality and the material world of physicality are more complimentary than oppositional in order to examine the space between the Internet and the installation that is emerging as small tech integrates human bodies and material ecologies at both local and global levels.

Miniaturization and the Virtual

In the ongoing explorations and attempts to engage with the effects of miniaturization and weightlessness, cultural theorists such as Pierre Lévy have focused on a parallel process called *virtualization*. In *Becoming Virtual: Reality in the Digital Age*, Lévy argues that digital technologies are part of a sea change in the "logic" of culture and society. Introducing this emergent logic, he criticizes the popular "misleading" opposition between virtuality and reality. Lévy writes, "[T]he word 'virtual' is often meant to signify the absence of existence, whereas 'reality' implies a material embodiment, a tangible presence" (23). This opposition, however, is an unproductive way of framing the value of virtual aspects of everyday life. Following the medieval derivation of the term (*virtualis* and *virtus*), Lévy reads "the virtual" as potentiality. "Strictly speaking," he concludes, "the virtual should not be compared with the real but the actual for virtuality and actuality are merely two different ways of being" (23). As digital technologies become increasingly smaller and recede into background networks that support communication and production, they generate the material grounds for potential human action, connection, and communication.

Opposed to actuality, the virtual is a problematic abstract concept. But as a network of material events, objects, and texts and their relations, it becomes a context for potential realities that might emerge from their nascent combination. If big physical things have already met their potential, materialized as they are in reality, virtual things are potential realities. Because these material possibilities exist as potentiality, one of the key characteristics of the virtual is "not there." According to Lévy, "When a person, community, act, or piece of information is virtualized, they are 'not there,' they de-territorialize themselves. A kind of clutch mechanism detaches them from conventional physical or geographical space and the temporality of the clock or calendar" (4). However, this "not there" is still an aspect of material reality. As Graham Harman sees it in *Guerrilla Metaphysics*, "any relation must count as a substance" (85). Objects are nothing more than densely interwoven sets of relations. As such, these sets both recede from human ability to have access to them and exert material force in the world. Likewise, as smaller technologies recede from human scale into the background, they seem as if they are not there, but they still exert considerable force on what becomes humanly possible.

Cultural theorists such as Walter Benjamin, Jean Baudrillard, and Paul Virilio have theorized virtual phenomena in ways that preface this new conception of the virtual. In "This Space for Rent," Benjamin foresees the dematerial qualities of the coming virtual media culture in the red neon that flickers on the street before his feet. The neon reflection is a protovirtual screen or space that, for Benjamin, signifies the virtual erasure of critical distance and the collapse of the subject into the media object ("One-Way Street" 476). Similarly, in the opening pages of *Fatal Strategies*, Baudrillard alludes to a reality in which (material) objects have broken free of their orbits around (human) subjects and

like Shaviro's remarks that everything a sort of reality (& M. Hansen re: computer interface = virtual reality) are transforming into a virtual mediascape. The spreading change in form or state from material to media object collapses subject and object equally into a virtual space. And in works such as *Open Sky*, Virilio writes of a world in which media-based phenomena act like quanta, existing in a virtual state of potential connectivity. Although these theorists carry an ambivalence toward virtual phenomena, their work provides useful conceptual starting places for moving toward a more material conception of the virtual. First, these works show the limitations of conventional humanities scholarship that focuses on the actual. Such scholarship remains locked in an either/or valuation of the virtual or the actual. But as Harman notes, the subject/object divide only exists at the level of human scale. Second, these works provide novel ways of valuing the virtual, even if the methods are infused with a sense of distrust. For example, Virilio's turn to modern physics allows him to characterize the virtual in ways that traditional linguistic and semiotic theory cannot. Although Virilio laments the rise of the virtual, arguing that humanity is defined by its resistances to the virtualization of space and time, his analogy to quantum physics provides a provocative way of articulating the emerging space in which small tech operates.

Just as contemporary theories of quantum physics are undoing the traditional, physical laws of Newtonian physics, the field of weightless economics allows for a revaluation of the virtual. Perhaps the simplest way to understand the "parallel process" of virtuality and actuality is in terms of a key property of weightless economics, its *infinite expansibility*. In her study of weightlessness, Coyle cites the work of Danny Quah: "Dematerialised commodities show no respect for space and geography. . . . Simply put, this means that the use of a dematerialised object by one person does not prevent another from using it. Other people can simultaneously use the word processing code I use as I type this. It is an economic good whose ownership cannot be transferred or traded, but simply replicated" (Coyle 3).

Objects in a weightless or virtualized culture are not bound by conventional rules of physical law, but this does not make them any less important as objects that exert force in the world. If anything, their affective capacity is enhanced. Weightless, virtualized objects, like software, can exist in more than one place simultaneously. This virtual reproducibility creates the potential for infinite material relations and actions. If nothing else, the peer-to-peer (P2P) downloading of music shows how miniaturized and weightless technologies create new conditions of possibility for relations that don't fit into traditional industrial economics. It shows the power of the network in relation to both the individuals who participate in it and the corporations who want to control it. What becomes so provocative about small tech is its move away from the traditional notion of virtual interface to new physical interfaces embedded into material ecologies.

From Interface to Environment

One of the key developments in contemporary digital culture is the shift in emphasis from software to hardware. Virtual miniaturization is not only made possible through software but also through advances in material devices. Objects such as cell phones, PDAs, and iPods constitute the popular dimensions of an emerging handheld culture. McLuhan might read these technologies as extensions of human bodies. But thinking in terms of Andre Leroi-Gourhan's revalorization of the hand in *Gesture and Speech* provides an important reconceptualization of handheld technologies. For him, prehistoric, prephonetic man communicates through sight, enacted through gesture, and through sound, enacted through speech. Gesture, however, gets separated from the human body with the development of cave paintings, and once phonetics is developed, the imagistic and gestural dimensions of speech become further removed from the body. Leroi-Gourhan argues that the twentieth century is producing a return to the techniques of the hand, but this return is post-McLuhan in its thinking. As Ariella Azoulay writes in her book *Death's Showcase*, "The hand may be part of a real body, but it is also an organ floating in the world to which the individual is extending a hand, responding to its interpellation" (12). Having been separated from the body through the development of writing, the hand is being (re)integrated into a new context. Rather than being an extension of the human body, the hand becomes a small technology in its own right, with the ability to enter into material combinations with other digital devices and material events to create new possibilities for communication and action. Consequently, the emerging handheld culture is signaling a move in emphasis from traditional, visual software interfaces to the ad hoc interfaces created by gestural objects in material ecologies.

In *Interface Culture*, Steven Johnson chronicles the rise of the personal computer and its transition into network culture in the 1990s that is made possible by the shift from line-code computing to the desktop interface. Like Leroi-Gourhan's externalization of gesture and speech into image and text, the contemporary move from a text-based interface to an image-based interface creates a new kind of information space. For Johnson, Simonedes's "memory palaces" from ancient Greek rhetoric are the first information spaces. Using imagined buildings as mnemonic devices plays on the mind's ability to remember images more easily than text (Johnson 12). The move from codes and commands to folders and windows plays on the same visual strength to create a new, more usable information space. The window and the mouse, invented by Douglas Engelbart in the mid- to late 1960s, allow the user to enter the world of the screen for active manipulation. Johnson argues that thinking of technology in terms of prosthetics (as the Italian futurists, McLuhan, Vannevar Bush, and even Engelbart at times did) is essentially an industrial perspective. Engelbart's interface initiates the move from tool as prosthetic extension to technology as environment, a space to be lived in and explored (24). Engelbart's windows, however, do not overlap, making the space essentially 2D. It is Alan Kay at Xerox PARC (Palo Alto Research Center) who, in the 1970s, extends the desktop metaphor so that various virtual pages can be layered. This more closely approximates 3D depth and makes it easier to see the virtual desktop as an environment (47). Steve Jobs, who saw a demo of the visual interface at PARC, ultimately popularizes the desktop metaphor with the spread of personal computers in the mid-1980s. But once PCs become networked in the 1990s, the simplicity of the desktop interface that contributes to its wide adoption gives way to a more complex information space and completes the transition from tool to virtual medium.

Johnson is correct to make the move from tool to space, but in *Interface Culture* he is still looking for a more complex metaphor or conscious representation of information space that can be built into a more complex interface. For Malcolm McCullough, on the other hand, digital culture is already in the midst of another shift that moves technologies from the virtual foreground into the material background. In *Digital Ground*, McCullough argues that computer technologies are moving from the explicit realm of the interface to the tacit architectures of houses, buildings, and environments. In the 1980s, the graphical user interface (GUI) is developed for consciously processing symbols on a stand-alone PC. It grabs the user's attention and monopolizes conscious manipulation of data stored locally or individually. The networking of individual PCs in the 1990s creates a move from interface to interaction. The Internet makes connected online environments and virtual social lives possible, but more software technologies are needed to handle the increasing amounts of information. These technologies become accumulated in the background of the PC, but they require even more interface attention to learn and operate them all. For McCullough, accumulating more software and applications for the desktop is not the answer. Users will end up with more applications than they can consciously learn to utilize.

Rather than continue to assume the need for universal connectivity—the need to be connected to all information all the time—McCullough is following current work in pervasive computing that focuses on embedding specific information technologies into the ambient backgrounds of local physical spaces. Building material ecologies instead of virtual environments separates information from a global network, places it in local contexts, and removes the pressure on conscious attention. Just as someone doesn't need to know a refrigerator runs on an electric motor in order to operate it, users shouldn't need to know the ins and outs of information appliances to use them. McCullough follows usability guru Don Norman in arguing that a tool should fit the task and situation so well that it goes unseen and remains a tacit part of use. As information technologies become a part of the ambient social infrastructure, they become parts of material situations, not just isolated tools with connections to the virtual network. Participating in a situation then, rather than interacting with others, operates on embodied, tacit knowledge where hardware is in the material background and freed from the PC.

Embedded technologies are not that mysterious or uncommon. Thermostats automatically heat and cool houses. Sensors detect E-ZPass tags in cars as they pass through tollbooths. Wireless technologies log on to the Internet through devices embedded in a wall or on a telephone pole. And microchips go into many more things than computers: the things people drive, carry, wear, and inhabit. Microprocessors embedded in everyday things from pocket radios to mobile phones pack more processing and memory capacity than a mainframe in the 1960s and initiate the drive toward miniaturization. Small sensors—in everything from smoke detectors to motion detectors to smog detectors—identify changing conditions in an environment and transmit the data to servers. Radio frequency ID tags can be placed in devices or products to heighten the ability of sensors. Like the UPC bar code, newer EPC (electronic product code) tags can create electronic codes for each individual product, not just a product type, and smart shelves can detect when something is taken off and then tabulate the data on product movement. Global positioning technologies, from maps in cars to wearable wrist bands for tracking children, can link up fixed and mobile sites, connecting place to data and information systems. Small tech is even being integrated into biological systems. In *Biomedia*, Eugene Thacker gives a number of examples: in vivo biosensors used for measuring blood pressure or releasing a drug into the bloodstream; handheld devices used for isolating, purifying, or amplifying DNA; microchips used for sequencing and analyzing proteins and genes (67–70). In all of these cases, the technologies can be connected to larger networks in order to increase the capacity to collect data and adapt to local conditions—like a thermostat or automatic updates on your computer, they can create feedback loops and alter a system's operation in response to local conditions.

For McCullough, our "growing constellations of devices" (69) or "ad hoc physical aggregations of digital devices" (70) are becoming systems themselves. And these systems do not just operate in the background on their own but can also interact with the human hand. Just as Johnson sees the shift from line code to visual interface as an advance to a simpler form of communication, McCullough sees the shift from typing code, clicking a mouse, or knowing computer programs and jargon to the use of speech and gesture to trigger embedded technologies as an advance. A tag can be placed in city locations to be triggered by a handheld PDA or cell phone and grant access to historical data, information for bar hopping, or navigation for tracking friends. Newer controls are beginning to move from industrial buttons to "haptic interface strategies based on gesture, glides, [or] motion sensing" (McCullough 85) that function through more responsive movements like joysticks on games. And even though haptics—technologies that interface with users via touch, vibrations, and/or motions—operate as a new form of gesture, text won't go away. It will still be ubiquitous, and our relationship to place will be greater as a wider variety of display surfaces will be emerging from PDAs to embedded digital signs (the text screen on a gas pump, for example). Virtual displays and touch screens can even be projected onto physical surfaces or into eyeglasses.

Connected to these constellations and aggregations, the hand is often separated from a conscious human intention and responds to its interpellation tacitly as its own independent object. In short, McCullough's work more concretely signals a move from the virtual as not real to the virtual as a tacit aspect of material reality and potentiality. This means that small tech is not just mobile but that handheld devices connect to fixed devices embedded in material contexts and participate in complex ecologies. Small tech enters into temporary ecological relations with embedded technologies and traditional interfaces.

Toward Methodologies for Small Tech

Johnson's primary argument in *Interface Culture* is that interface design is just as important as literature and thus deserves its own form of criticism on par with literary criticism that could combine art and technology. In the light of McCullough's argument, however, the key is not to develop a new criticism for interface designs but new methodologies for mapping these more complex ecologies among the interface, the hand, and the built environment. Since the world is already in the process of being layered with digital systems, theorists and designers have to embrace material place as well as virtual interface. The problem, according to McCullough, is that currently no methodologies exist for this difficult task. And understandably so. With the shift from one global network to innumerable local ecologies that may or may not be connected to this network, a single method or even a few methodologies would be insufficient for the task. It seems more likely that a theorist or designer would have to assess local conditions and build ad hoc methods for understanding and living in that particular configuration. Rather than applying methods deductively, multiple methods would need to be remixed in the context of specific ecologies. Looking back to certain theoretical precursors for some conceptual starting places would be a first step toward generating these context-specific methodologies and including humans along with art and technology as elements of concern.

While McLuhan, with Questin Fiore in *War and Peace in the Global Village*, is caught in a particular time and place with respect to technologies, which places certain limits on his perspective, he is one of the first critics to look at technology as an environment and invent a method for assessing its effects and affects on human culture. "The important thing," he writes, "is to realize that electronic information systems are live environments in the full organic sense. They alter our feelings and sensibilities, especially when they are not attended to" (36). For McLuhan, the shift from industrial tools to electronic environments produces a shift from figure (content) to ground (medium), from mind to body, from logic to affect, from the textual to the tactile, and from extended print texts to fragmented and collaged image-texts. In order to bring these effects to the foreground, McLuhan posits four general laws of technology. In his analysis, all media generate four kinds of effects, which he turns into heuristic questions: What does the medium amplify? What does it obsolesce? What does it retrieve? What does it reverse? McLuhan, with Eric McLuhan in *Laws of Media: The New Science*, points out that these laws aren't meant as separate and static, but they reveal a dynamic pattern of interlocking effects: "[The] tetrad of the effects of technologies and artifacts presents not a sequential process, but rather four simultaneous ones. All four aspects are inherent in each artifact from the start" (99). This recognition situates McLuhan as attempting to derive a method that addresses the complex relationship between a technology and its surrounding cultural and material ecology.

Despite his clear move from isolated figure to complex ground, McLuhan is still caught up in the desire to derive general scientific principles and a somewhat deterministic cause-and-effect logic. He is still operating at an abstract level and looking at the necessary effects of specific technologies rather than the shifting effects of ecologies. Martin Heidegger, on the other hand, though he slightly predates McLuhan, is looking at particular technologies in particular configurations or constellations, which is more appropriate for examining small tech and the kinds of aggregates that McCullough sees on the horizon. As small tech becomes a more significant aspect of everyday life, increasingly becoming smaller and more mobile, it is becoming more and more difficult to see technology as producing a particular set of effects or an all-encompassing media environment like the large-scale TV culture that McLuhan encountered. More contemporary, complex ecologies require methods that are increasingly specific to their immediate fields of relations. Heidegger's phenomenological method is after the conditions of possibility that a particular tool in a particular environment establishes. His premise is that specific modes of being come alive or show up in certain ecological constellations. Any technology is always experienced in terms of its belonging to other equipment: "ink-stand, pen, ink, paper, blotting pad, table, lamps, furniture, windows, doors, room" (Heidegger 97). Technologies, especially in the case of small tech, are never distinct objects: they are only experienced in relation to other entities arranged in complex constellations to form particular environments. What human bodies encounter is the room in its totality, and the room is not simply a geometrical space but is experienced ecologically "as equipment for residing" (Heidegger 98). This overall constellation sets the conditions of possibility or virtual potentiality for a particular act of writing.

Heidegger's method looks to examine one of two things: (1) these ecological arrangements can be drawn narrowly to determine a tool's specific manipulability or (2) they can be drawn more broadly to discern their manifold assignments. What shows up in the use of a technology is its "specific manipulability," a specific mode of being and the future structural possibility integral to it. A hammer's specific manipulability allows it to hammer a nail into a board in a way that a drinking glass doesn't. But a hammer (or any small tech) can only carry out its full potential within a larger ecological structure. Within larger ecological arrangements there is what Heidegger calls an "assignment." These assignments are the potential paths of future action and development that the whole ecology makes possible. A hammer's specific manipulability may allow it to hammer in a nail, but a hammer by itself doesn't carry the possibility, or assignment, for a house in the way a hammer, saw, nails, wood, a plane, a blueprint, and a human body carry the conditions of possibility for a house. Only a larger constellation can play out or exhibit the possible effects set up in its field of relations. Small technologies only emerge for human concern—the tacit uses the human body enacts through particular tools in particular worlds—out of such ecological arrangements. It is only through a constellation's relation to human bodies (or in the case of the hammer and many small technologies, the human hand) that the tool, its specific manipulability, its ecologies, and its manifold assignments can be discovered or known.

While McLuhan is a clear precursor to McCullough in that he is interested in the shift from figure to ground (McCullough 71), McCullough is more interested in the production of complex ecologies. In this sense, his ideas are more along the lines of Heidegger. Though Heidegger's famous tool analysis seems to center on the tool, the tool is really just a methodological starting point for mapping out ecological relations and conditions of possibility. Like McLuhan, Heidegger makes a move from center to periphery, but rather than technology causing effects, it combines with many other elements to create conditions of possibility that suggest potential futures rather than determine them. This combination of actual materiality and virtual potentiality is central to any discussion of methodology and small tech. In our current weightless context, as technologies get smaller and smaller, their materiality is less obvious and their potentiality is more powerful. Consequently, their forms of analysis need to focus on tighter, more detailed constellations and their specific conditions of possibility.

In *Writing Machines*, for example, N. Katherine Hayles looks at the ecologies around specific texts and puts forth media-specific analysis as a corresponding method of criticism. For Hayles, all texts must be embodied to exist and each specific embodiment affects linguistic, semiotic, and rhetorical elements of the work. The physicality of industrial technologies, however, implies that the specific properties of a medium, such as the link in hypertext, would function the same way across all hypertexts. But media-specific analyses would recognize that any work's materiality will be different with each ecology, enaction, and user. A media-specific analysis would set out (1) to account for the specificity of each particular engagement with a text and (2) to examine the way a particular embodiment can function to connect the user to a distributed network that communicates through the unconscious bodily experience of the medium. For example, Hayles discusses a student installation project that features a kiosk that literally invites a user to physically interface with the installation and participate in producing the text. By choosing links on the computer screen, the user produces images and words projected on the wall and printed out on an accompanying printer. In this case, the link creates different textual and material effects with each embodiment—each user will form an embodied relationship to the material context of the installation to produce a text particular to that constellation. For Hayles, the project "create[s] new sensory, physical, and metaphysical relationships between user and database" (*Writing Machines* 102). These relationships highlight more than the physicality of location. They make the materiality as well as the virtuality of the medial ecology show up. A media-specific analysis, then, would set out the specific elements in the particular ecology and examine the "interplay between form, content, and medium" (*Writing Machines* 31) that they make possible.

While Heidegger sees tools and ecologies in relation to human concern, Hayles locates human thought and action in distributed cognitive environments where humans are spliced into material-informational circuits. The navigation of a ship, for example, requires the distribution of decision-making across a complex interrelationship of technology, humans, and nature. This is a posthuman condition in which the human hand in an ecological formation carries more cognitive and material power than an isolated human body. As Hayles puts it, "Modern humans are capable of more sophisticated cognition than cavemen not because moderns are smarter . . . but because they have constructed smarter environments in which to work" (*How We Became Posthuman* 289). The development of smaller technologies and their integration into such ecologies is currently producing even smarter environments in which to work, and the layering of multiple distributed networks is producing the ambient digital ground in which humans live. If scholars in the humanities fail to address this level of complexity, their analyses of technologies will be caught up in applying an industrial methodology to a postindustrial, weightless world. The complex world of small tech means that multiple domains will come together in any given constellation, so multidisciplinary methods will have to be remixed to deal with such specific situations.

Mapping Complex Ecologies

In the light of this methodological complexity, *Small Tech* brings together scholars from rhetoric and literacy studies, digital and visual arts, and social and political sciences to assess the changing relationships among the traditional interface and digital networking and the emerging shift to mobility in relation to environment. In an attempt to understand this contemporary space of everyday life between cyberculture and new media, the authors look to remix disciplinary methods in order to map the affects, forms of work, products, actions, or relations that a small technology as a part of an ecological constellation can change or make possible. The mix of disciplinary approaches means that many of these mappings will differ in objective and outcome. As Deleuze and Guattari write, "The map is open and connectable in all of its dimensions; it is detachable, reversible, susceptible to constant modification. It can be torn, reversed, adapted to any kind of mounting, reworked by an individual, group, or social formation. It can be drawn on a wall, conceived of as a work of art, constructed as a political action or as a meditation" (12).

Such openness means that no single or predetermined approach can take center stage. Jay David Bolter and Richard Grusin take a semiecological approach in *Remediation*, looking at various technologies in relation to other technologies to show *a particular kind of relation* among them—remediation. *Small Tech*, however, has no such specific thesis or relation to show. Each author investigates the ecologies around a specific technology to make its potentiality show up, to chart its paths, to uncover its specific manipulabilities, to reveal its tacit knowledge, or to discover its possible assignment contexts. Consequently, each mix of methodologies and the map that emerges from their enaction will vary with the particulars of the specific technology and constellation in view.

One of the variations that might affect the outcome of a particular analysis is scale. A constellation could be drawn as large as the Internet or might zero in on something as small as digital code. For example, Tim Berners-Lee took advantage of the network already in place in order to invent the World Wide Web. By assessing the conditions of possibility in the Internet's constellation of nodes and wires, he was able to write new code to make new forms of textuality, work, and distribution possible among his physics and engineering colleagues at CERN (European Organization for Nuclear Research). But as importantly, Marc Andreesen, a student and part-time assistant at the National Center for Supercomputing Applications (NCSA) at the University of Illinois, made an almost more startling contribution to this emerging ecology. In the development of the Mosaic browser, Andreesen included an image code. Even though this inclusion troubled Berners-Lee because he thought it undermined his initial intentions for the Web, this one simple code opened the medium to a wealth of new possibilities and affects. When inserted into the larger network, the one code created an entirely new ambient environment and a flurry of activity to fill the ecological niches it made possible.

The long-term effect of this single code allowed Gregory Ulmer to take a large-scale view of the effects of this weightless, networked environment. For him, the significance of the Web is that it allows humans to manipulate and circulate images in a manner that is unprecedented in human history. The Web extends the juxtaposition and manipulation of images exhibited in works such as Andy Warhol's and consequently "locate[s] the existence of a new unit of meaning, in the way that the Greeks noticed the abstract idea (*dike*) once the story could be scanned visually" (Ulmer 167–68). The shift from orality to literacy allowed the Greeks to see common traits among characters that they never noticed during oral performances. Plato capitalized on this shift to written language by abstracting concepts and definitions from the texts—shifting from the actions of the characters to the abstract concepts they exhibited such as justice or virtue. Ulmer argues that the contemporary shift from literacy to electracy is similarly making a new way of seeing possible with respect to images. In addition to literal and figurative meanings, both of which appear in literacy, electracy foregrounds personal, associative meanings that create specific identifications with images and produce a mood or atmosphere particular to that constellation. In other words, rather than a global village in which all humans live, the small technologies that make up the Internet and make the pervasive distribution of images possible create specific constellations around singular bodies.

Many more possibilities can arise out of the current interface between the hand, the desktop, the network, and the environment and set the stage for analyses of traditional software in the context of new relations; investigations of small, mobile technologies and their affects; and examinations of newer technologies and their relation to ambient environments. Whether the discussion revolves around the scale of action scripts, Web logs, and cut-and-paste tools; the scale of iPods, PDAs, and cell phones; or the scale of haptics, wearable technologies, and virtual environments, each of the authors in *Small Tech* provides an analysis of a specific technology and its specific ecology, raises social or cultural issues about the affects and effects of this tool-context, and provides a set of concepts for thinking about the small technologies and their ecologies in new ways.

In the first section of chapters, "Traditional Software in New Ecologies," the authors investigate the ways in which a number of traditional software applications are transformed by emerging media ecologies. In the opening chapter, "Data Visualization as New Abstraction and as Anti-Sublime," Lev Manovich introduces the theory and practice of *dynamic data visualization* and argues for a revaluation of the conventional uses of mapping technologies. Extending the focus on imaging, Adrian Miles argues in "Softvideography" that digital video is going through a paradigm shift that echoes the shift from print to hypertext theorized by Jay David Bolter, Michael Joyce, and others in the 1990s. Miles writes, "The rise of domestic audiovisual technologies, for example software suites such as Apple's 'iLife' quartet (iDVD, iPhoto, iTunes, and iMovie), or Microsoft's Windows Movie Maker, threatens to do for home video what the original Macintosh achieved for the word with the invention of desktop publishing: the rise of desktop video." In "Technopolitics, Blogs, and Emergent Media Ecologies," Richard Kahn and Douglas Kellner widen their scope to demonstrate ways that blogs have made a new age of political activism possible. Introducing new software into the World Wide Web creates new possibilities for connection and political action.

In chapters by Karla Saari Kitalong and Lance Strate, each writer again describes the traditional software functions of layering and cutting and pasting respectively and explores the cultural effects of these practices. Sean D. Williams argues that WYSIWYG interfaces encourage a superficial engagement with the complex ecologies found on the World Wide Web and limit the designer's understanding of the medium to one plane of complexity. Collin Gifford Brooke examines the ulterior view that the traditional notion of linking in hypertext can perform new functions within the context of Macromedia Flash and create more textually complex environments. And in "ScriptedWriting," David Rieder explores the ways in which the ActionScript programming language can be used as a generative environment for user-driven writing projects, showing how work with code can produce more complex ecologies.

The second section, "Small Tech and Cultural Contexts," consists of shorter essays that investigate the ecologies surrounding a particular small technology and the cultural context that it affects. In "Overhearing: The Intimate Life of Cell Phones," Jenny Edbauer Rice redefines the cell phone as a technology that blurs the conventional boundaries between the public and the private to create a new affective ecology. With a related theme of boundary-blurring, Paul Cesarini forecasts the ways in which podcasting and the networked iPod have created the conditions of possibility for everyone to become a DJ on his or her own radio station. In "Walking with Texts," Jason Swarts describes the way PDAs transform organizing and reading texts on-screen. And Wendy Warren Austin reads cell phone text-messaging and Internet messaging as practices that recontextualize communication and connectivity.

Just as hardware reconfigures these complex ecologies, the software that connects the virtual to the actual also plays a central role. Michael Pennell maps the ways peer-to-peer software and the networking it makes possible are transforming the cultural context for identity formation. Comparing Google and Yahoo, Johndan Johnson-Eilola argues that the "postmodern space" created by Google creates the potential for new forms of networking. Returning to hardware in "Let There Be Light in the Digital Darkroom," Robert A. Emmons Jr. examines how digital cameras have produced a completely new ecology surrounding images. Similarly, Veronique Chance explores the way digital video cameras change the ecology surrounding live performance. The theme of shifting contexts continues with Julian Oliver's "Buffering Bergson," which argues that the increasingly generative aspects of user-driven computer games need a theory of dynamism such as Bergson's to account for the affects of emergence in gaming environments. And finally, Teri Rueb examines the dynamic aspect of global positioning software (GPS) technologies and the effects it can have on point-of-view in narrative performances.

In the final section of chapters, "Future Technologies and Ambient Environments," the authors describe a number of futures for virtual technologies, many of which are emerging now. In "Virtual Reality as a Teaching Tool," James J. Sosnoski describes his work with the "Virtual Harlem" project, a virtualized teaching environment in which students can experience historically accurate neighborhoods that have since been razed. Johanna Drucker also discusses a pedagogical environment in "Digital Provocations and Applied Aesthetics." The goal of the speculative computing she describes is to place a student body in the context of a digital game to enhance learning at the bodily, affective level. Isabel Pedersen shifts the focus on the body from the pedagogical context to the social world. She explores the emergent technology of wearable computers to show how mobile technologies are being designed by artists to augment personal experience and redefine everyday life. In the chapter "Sousveillance," Jason Nolan, Steve Mann, and Barry Wellman document their work developing wearable technologies that interrogate the surveillance culture in which we live by surveilling the surveillers. Such experimental work forecasts the emergence of handheld technology and its affects on immediate cultural ecologies.

The next two chapters turn from social contexts toward the home. Jim Bizzocchi, in "Ambient Video," discusses the ways digital video is serving to create new ambient environments in the domestic sphere. And in "Sound in Domestic Virtual Environments," Jeremy Yuille extends this logic to the way sound functions in video games to create a tacit environment that greatly affects game action. In her chapter, "Getting Real and Feeling in Control," Joanna Castner Post challenges readers to recognize the historical limits of the desktop metaphor and to imagine the future of haptic, tactile technologies. Provocatively, Castner Post argues that the linguistic process of naming things will change when tactility is introduced into the process. Mark Paterson extends the discussion of haptics to the combination of visual display and haptic interface. This marriage could create wholly new virtual environments that are faithful to the properties and sensations of material objects. Such developments could signal the future importance of gesture and the floating hand in ways that parallel the importance of image manipulation in contemporary digital culture.

Works Cited

"About Small Tech." *Michigan Small Tech: Growing the Micro and Nano Industry.* 12 May 2005.
  http://www.michigansmalltech.com/AboutSmallTech.
Azoulay, Ariella. *Death's Showcase: The Power of Image in Contemporary Democracy.* Cambridge, Mass.: MIT
  Press, 2003.

Baudrillard, Jean. *Fatal Strategies*. Trans. Philip Beitchman and W. G. J. Niesluchowski. Ed. Jim Fleming. Brooklyn: Semiotext(e), 1990.

Benjamin, Walter. "One-Way Street." *Walter Benjamin, Selected Writings. Vol. I: 1913–1926*. Trans. Edmund Jephcott. Ed. Marcus Bullock and Michael Jennings. Cambridge, Mass.: Belknap Press, Harvard University Press, 1996. 444–88.

Bolter, Jay David, and Richard Grusin. *Remediation: Understanding New Media*. Cambridge, Mass.: MIT Press, 1999.

Coyle, Diane. *The Weightless World: Strategies for Managing the Digital Economy*. Cambridge, Mass.: MIT Press, 1999.

Deleuze, Gilles, and Félix Guattari. *A Thousand Plateaus: Capitalism and Schizophrenia*. Trans. Brian Massumi. Minneapolis: University of Minnesota Press, 1987.

Doyle, Richard. *Wetwares: Experiments in Postvital Living*. Minneapolis: University of Minnesota Press, 2003.

Greenspan, Alan. "Remarks by Chairman Alan Greenspan: Technological Advances and Productivity." *Federal Reserve Board*. 16 Oct. 1996. http://www.federalreserve.gov/boarddocs/speeches/1996/19961016 .htm.

Harman, Graham. *Guerrilla Metaphysics: Phenomenology and the Carpentry of Things*. Chicago: Open Court, 2005.

Hayles, N. Katherine. *How We Became Posthuman: Virtual Bodies in Cybernetics, Literature, and Informatics*. Chicago: University of Chicago Press, 1999.

———. *Writing Machines*. Cambridge, Mass.: MIT Press, 2002.

Heidegger, Martin. *Being and Time*. Trans. J. Macquarrie and E. Robinson. San Francisco: Harper Collins, 1962.

Johnson, Steven. *Interface Culture: How New Technology Transforms the Way We Create and Communicate*. San Francisco: Harper Collins, 1997.

Leroi-Gourhan, Andre. *Gesture and Speech*. Trans. Anna Bostock Berger. Cambridge, Mass.: MIT Press, 1993.

Lévy, Pierre. *Becoming Virtual: Reality in the Digital Age*. Trans. Robert Bononno. New York: Plenum Press, 1998.

Manovich, Lev. *The Language of New Media*. Cambridge, Mass.: MIT Press, 2001.

———. "New Media from Borges to HTML." Introduction. *The New Media Reader*. Ed. N. Wardrip-Fruin and N. Montfort. Cambridge, Mass.: MIT Press, 2003. 13–25.

McCullough, Malcolm. *Digital Ground: Architecture, Pervasive Computing, and Environmental Knowing*. Cambridge, Mass.: MIT Press, 2004.

McLuhan, Marshall, and Eric McLuhan. *Laws of Media: The New Science*. 1988. Toronto: University of Toronto Press, 1992.

McLuhan, Marshall, and Questin Fiore. *War and Peace in the Global Village*. 1968. San Francisco: HardWired, 1997.

Rifkin, Jeremy. *The Age of Access: The New Age of Hypercapitalism, Where All of Life Is a Paid-for-Experience*. New York: Jeremy Press. Tarcher/Putnam, 2000.

Shaviro, Steven. *Connected: or, What it Means to Live in the Network Society*. Minneapolis: University of Minnesota Press, 2003.

Thacker, Eugene. *Biomedia*. Minneapolis: University of Minnesota Press, 2004.

Ulmer, Gregory. *Internet Invention: From Literacy to Electracy*. New York: Longman, 2003.

Virilio, Paul. *Open Sky*. Trans. Julie Rose. London: Verso, 1997.

Traditional Software in New Ecologies

1

Data Visualization as New Abstraction and as Anti-Sublime

Lev Manovich

Visualization and Mapping

Along with a graphical user interface, a database, navigable space, and simulation, *dynamic data visualization* is one of the genuinely new cultural forms enabled by computing. Of course, the fans of Edward Tufte will recall that it is possible to find examples of graphical representation of quantitative data already in the eighteenth century, but the use of the computer medium turns such representations from the exception into the norm. It also makes possible a variety of new visualization techniques and uses for visualization. With computers we can visualize much larger data sets; create dynamic (i.e., animated and interactive) visualizations; feed in real-time data; base graphical representations of data on mathematical analysis using a variety of methods from classical statistics to data mining; and map one type of representation onto another (e.g., images onto sounds, sounds onto 3D spaces).

Since René Descartes introduced the system for quantifying space in the seventeenth century, graphical representation of functions has been the cornerstone of modern mathematics. In the last few decades, the use of computers for visualization enabled development of a number of new scientific paradigms such as chaos and complexity theories and artificial life. It also forms the basis of a new field of scientific visualization. Modern medicine relies on visualization of the body and its functioning; modern biology similarly is dependent on visualization of DNA and proteins. But while contemporary pure and applied sciences, from mathematics and physics to biology and medicine, heavily rely on data visualization, in the cultural sphere visualization until recently has been used on a much more limited scale, being confined to 2D graphs and charts in the financial section of a newspaper or an occasional 3D visualization on television to illustrate the trajectory of a space station or of a missile.

I will use the term *visualization* for situations in which quantified data *are not visual*— the output of meteorological sensors, stock market behaviors, the set of addresses describing the trajectory of a message through a computer network, and so on—and are transformed into a visual representation.[1] The concept of *mapping* is closely related to visualization, but it makes sense to keep it separate. By representing all data using the same numerical code, computers make it easy to map one representation onto another: grayscale images onto a 3D surface, a sound wave onto an image (think of visualizers in music players such as iTunes), and so on. Visualization then can be thought of as a particular subset of mapping in which a data set is turned into an image.

Human culture practically never uses more than four dimensions in its representations because we humans live in 4D space. Therefore we have difficulty imagining data in more than these four dimensions: three dimensions of space (X, Y, Z) and time. However, more often than not, the data sets we want to represent have more than four dimensions. In such situations, designers and their clients have to choose which dimensions to use, which to omit, and how to render the selected dimensions.

This is the new politics of mapping in computer culture. Who has the power to decide what kind of mapping to use, what dimensions are selected, what kind of interface is provided for the user? These new questions about data mapping are now as important as more traditional questions about the politics of media representation by now well rehearsed in cultural criticism (e.g., who is represented and how, who is omitted). More precisely, these new questions about the politics of quantified data representation run parallel to the questions about the content of the iconic and narrative media representations. In the latter case, we usually deal with the visual images of people, countries, and ethnicities; in the former case, the images are abstract 3D animations, 3D charts, graphs, and other types of visual representation used for quantified data.

Data Modernism

Mapping one data set into another, or one media into another, is one of the most common operations in computer culture, and it is also common in new media art.[2] One of the earliest mapping projects, which received lots of attention and lies at the intersection of science and art (because it seems to function well in both contexts), was Natalie Jeremijenko's "live wire." Working in Xerox PARC (Palo Alto Research Center) in the early 1990s, Jeremijenko created a functional wire sculpture that reacts in real time to network behavior: more traffic causes the wire to vibrate more strongly. In the last few years, data mapping has emerged as one of the most important and interesting areas in new media art, attracting the energy of some of the best people in the field. It is not accidental that out of ten net art projects included in the 2002 Whitney Biennale, about a half presented different kinds of mapping: a visual map of the space of Internet addresses (Lisa Jevbratt), a 3D navigable model of Earth presenting a range of information about the earth in multiple layers (John Klima), another 3D model illustrating the algorithm used for genome searches (Benjamin Fry), and diagrams of corporate power relationships in the United States (John On and Futurefarmers) (*Whitney ArtPort*).

Let's discuss a couple of well-known data visualization art projects in more detail. In her project *1:1*, Lisa Jevbratt created a dynamic database containing IP (Internet protocol) addresses for all the hosts on the World Wide Web, along with five different ways to visualize this information. As the project description by Jevratt points out: "When navigating the web through the database, one experiences a very different web than when navigating it with the 'road maps' provided by search engines and portals. Instead of advertisements, pornography, and pictures of people's pets, this web is an abundance of non-accessible information, undeveloped sites, and cryptic messages intended for someone else. . . . The interfaces/visualizations are not maps of the web but are, in some sense, the web. They are super-realistic and yet function in ways images could not function in any other environment or time. They are a new kind of image of the web and they are a new kind of image."

In a 2001 project *Mapping the Web Infome*, Jevbratt continues to work with databases, data gathering, and data visualization tools, and she again focuses on the Web as the most interesting data depository corpus available today. For this project, Jevbratt wrote special software that enables easy menu-based creation of Web crawlers and visualization of the collected data (a crawler is a computer program that automatically moves from Web site to Web site collecting data from them). She then invited a number of artists to use this software to create their own crawlers and also to visualize the collected data in different ways. This project exemplifies a new function of an artist as a designer of software environments that are then made available to others.

Alex Galloway and the RSG (Radical Software Group) collective uses a similar approach in his network visualization project *Carnivore* (2002). Like Jevbratt, Galloway and the RSG collective created a software system that they opened up to other artists to use. Physically, *Carnivore* was at some point styled like a morph between a nondistinct box for telephone surveillance, such as the ones used in GDR, and a modernist sculpture. Connected to some point in the network, it intercepts all data going through it. This by itself does not make it art, because a number of commercial software packages perform similar functions. For instance, Etherpeek 4.1 is a LAN (local area network) analyzer that captures packets from attached Ethernet or AirPort networks and uses decoding to break these packets into their component fields. It can decode FTP, HTTP, POP, IMAP, Telnet, Napster, and hundreds of other network protocols. It performs real-time statistical analysis of captured packets and can reconstruct complete e-mail messages from the captured packets.

As is often the case with artist software, *Carnivore* only offers a small fraction of the capabilities of its commercial counterparts such as Etherpeek. What it does offer instead is the open architecture that allows other artists to write their own visualization clients that display the intercepted data in a variety of different ways. Some of the most talented artists working with the Net have written visualization clients for *Carnivore*. The result is a diverse and rich menu of forms, all driven by the network data.

Having looked at some classical examples of data visualization art, I would like now to propose a particular interpretation of this practice by comparing it with early twentieth century abstraction. In the first decades of the twentieth century, modernist artists mapped the *visual chaos* of the metropolitan experience into simple geometric images. Similarly, we can say data visualization artists transform the *informational chaos* of data packets moving through the network into clear and orderly forms.

In addition, we can make another parallel. Modernism reduced the particular to its Platonic schemas (think of Piet Mondrian, for instance, systematically abstracting the image of a tree in a series of paintings). Similarly, data visualization is engaged in a reduction as it allows us to see patterns and structures behind the vast and seemingly random data sets. Thus it is possible to think of data visualization as a new abstraction.

This parallel should be immediately qualified by noting some of the crucial differences. Modernist abstraction was in some sense antivisual—reducing the diversity of familiar everyday visual experience to highly minimal and repetitive structures (again, Mondrian's art provides a good example). Data visualization often employs the opposite strategy: the same data set drives endless variations of images (think of various visualization plug-ins available for music players such as iTunes, or *Carnivore*'s clients). Thus *data visualization moves from the concrete to the abstract and then again to the concrete*. The quantitative data is reduced to its patterns and structures that are then exploded into many rich and concrete visual images.

Another important difference is the new quality of many data visualizations that can be called *reversibility*. After Mondrian, Robert Delaunay, Pablo Picasso, and other modernist artists reduced the concrete sensible reality to abstract schemes shown in their paintings, the viewer cannot get back to reality by clicking on the painting. In other words, the reduction only operates in one direction. But with many data visualization images, the user can interact with a visualization to get more information about the data that gave rise to these images, evoke other representations of these data, or simply access the data directly. A good example of this is the elegant and evocative visualization *Anemone* by one of the masters of data visualization, Benjamin Fry. The visualization presents an organic-looking, constantly growing structure that is driven by the structure of a particular Web site and the access statistics of this site. In the default view, no labels or text accompany the visualization, so a still from the visualization can be at first sight mistaken for a typical modern abstract painting in the genre of "organic abstraction." However, at any time the user can click on any part of the moving structure to reveal the labels that explain which data is being represented by this part (in this case, a particular directory of the Web site). So what may be mistaken as the pure result of an artist's imagination is in fact a precise map of the data. In short, the visualization is "reversible"—it allows its user to get back to the data that gave rise to this visualization.

Meaningful Beauty: Data Mapping as Anti-sublime

Along with relating data visualization to modernist abstraction, let me now sketch another interpretation that will connect it to another concept in modern cultural history—that of the sublime. Since the sublime is too often and maybe too easily invoked in relation to various spectacular phenomena of contemporary culture, I will immediately negate this term. My argument goes as follows.

Data visualization projects often carry the promise of rendering a phenomenon that is beyond the scale of human senses as something that is within our reach, something visible and tangible. It is not, therefore, by accident that the most celebrated and admired examples of data visualizations (especially as used in science) show the structures on superhuman scale both in space and time: the Internet, astronomical objects, geological formations developing over time, global weather patterns, and so on. This promise makes data mapping into the exact opposite of Romantic art concerned with the sublime. In contrast, data visualization art is concerned with the *anti-sublime*. If Romantic artists thought of certain phenomena and effects as unrepresentable, as going beyond the limits of human senses and reason, data visualization artists aim at precisely the opposite: to map such phenomena into a representation whose scale is comparable to the scales of human perception and cognition. For instance, Jevbratt's *1:1* reduces cyberspace—usually imagined as vast and maybe even infinite—to a single image that fits within the browser frame. Similarly, the graphical clients for *Carnivore* transform another invisible and "messy" phenomenon—the flow through the network of data packets that belong to different messages and files—into ordered and harmonious geometric images. The macro and the micro, the infinite and the finite are mapped into manageable visual objects that fit within a single browser frame.

The desire to take what normally falls outside of the scale of human senses and to make it visible and manageable aligns data visualization art with modern science. Its subject matter (i.e., data) puts it within the paradigm of modern art. In the beginning of the twentieth century, art largely abandoned one of its key (if not *the* key) functions—portraying the human being. Instead, most artists turned to other subjects, such as abstraction, industrial objects and materials (e.g., Marcel Duchamp, minimalists), media images (pop art), the figure of the artist herself or himself (performance and video art)—and now data. Of course, it can be argued that data visualization represents the human being indirectly by visualizing her or his activities (typically the movements through the Net). Yet, more often than not, the subjects of data visualization projects are objective structures (such as the typology of Internet) rather than the direct traces of human activities.

Motivation Problem

As I already noted in the beginning, it is possible to think of visualization practices as a particular example of the more general operation digital computers are very good at: mapping. The relatively easy way in which we can use computers to make any set of data in any medium into something else opens up all kinds of opportunities but also creates a new kind of cultural responsibility.

We can rephrase this issue by talking about arbitrary versus motivated choices in mapping. Since computers allow us to easily map any data set into another set, I often wonder, why did the artist choose this or that form of visualization or mapping when endless other choices were also possible? Even the very best works that use mapping suffer from this fundamental problem. This is the "dark side" of the operation of mapping and of computer media in general—its built-in existential angst. By allowing us to map anything onto anything else, to construct an infinite number of different interfaces to a media object, to follow infinite trajectories through the object, and so on, computer media simultaneously make all these choices appear arbitrary—unless the artist uses special strategies to motivate her or his choices.

Let's look at one example of this problem. One of the most outstanding architectural buildings of the last two decades is the Jewish Museum in Berlin by Daniel Libeskind. The architect put together a map that showed the addresses of Jews who were living in the neighborhood of the museum site before World War II. He then connected different points on the map together and projected the resulting net onto the surfaces of the building. The intersections of the net projection and the design became multiple irregular windows. Cutting through the walls and the ceilings at different angles, the windows point to many visual references: the narrow eyepiece of a tank; the windows of a medieval cathedral; exploded forms of the cubist, abstract supermatist paintings of the 1910s–1920s. Just as in the case of Janet Cardiff's audio tours that weave fictional narratives with information about the actual landscape, here the virtual becomes a powerful force that reshapes the physical. In the Jewish Museum, the past literally cuts into the present. Rather than something ephemeral, here data space is materialized, becoming a sort of monumental sculpture.

But there was one problem that I kept thinking about when I visited the still-empty museum building in 1999—the problem of motivation. On the one hand, Libeskind's procedure to find the addresses, make a map, and connect all the lines appears very rational, almost the work of a scientist. On the other hand, as far as I know, he does not tell us anything about why he projected the net in this way as opposed to any other way. So I find something contradictory in the fact that all the painstakingly collected and organized data at the end is arbitrarily "thrown" over the shapes of the building. I think this example illustrates well the basic problem of the whole mapping paradigm. Since usually there are endless ways to map one data set onto another, the particular visualization chosen by the artist often is unmotivated, and as a result, the work feels arbitrary. We are always told that in good art "form and content form a single whole" and that "content motivates form." Maybe in a "good" work of data art, the map used would have to somehow relate to the content and context of data—or maybe this is the old criterion that needs to be replaced by some new one.

One way to deal with this problem of motivation is to not to hide but to foreground the arbitrary nature of the chosen mapping. Rather than try to always be rational, data visualization art can instead make the method out of irrationality.[3] This, of course, was the key strategy of the twentieth century Surrealists. In the 1960s, the late surrealists—the situationists—developed a number of methods for "the dérive" (the drift). The goal of "the dérive" was a kind of spatial "ostranenie" (estrangement): to let the city dweller experience the city in a new way and thus politicize her or his perception of the habitat. One of these methods was to navigate through Paris using a map of London. This is the kind of poetry and conceptual elegance I find missing from mapping projects in new media art. Most often these projects are driven by the rational impulse to make sense out of our complex world, a world in which many processes and forces are invisible and are out of our reach. The typical strategy then is to take some data set—Internet traffic, market indicators, Amazon.com book recommendations, or weather—and map it in some way. This strategy echoes not the aesthetics of the surrealists but a rather different paradigm of the 1920s leftist avant-garde. The similar impulse to "read off" underlying social relations from the visible reality animated many leftist artists in the 1920s, including the main hero of my book *The Language of New Media*—Dziga Vertov. Vertov's 1929 film *A Man With a Movie Camera* is a brave attempt at visual epistemology—to reinterpret the often banal and seemingly insignificant images of everyday life as the result of the struggle between the old and the new.

Important as the data visualization and mapping projects are, they seem to miss something else. While modern art tried to play the role of "data epistemology," thus entering into competition with science and mass media to explain to us the patterns behind all the data surrounding us, it also always played a more unique role: to show us other realities embedded in our own, to show us the ambiguity always present in our perception and experience, to show us what we normally don't notice or don't pay attention to. Traditional "representational" forms—literature, painting, photography, and cinema—played this role very well.

For me, the real challenge of data art is *not* about how to map some abstract and impersonal data into something meaningful and beautiful—economists, graphic designers, and scientists are already doing this quite well. The more interesting and, in the end, maybe more important challenge is how to represent the personal subjective experience of a person living in a data society. If daily interaction with volumes of data and numerous messages is part of our new "data-subjectivity," how can we represent this experience in new ways? How can new media represent the ambiguity, the otherness, and the multidimensionality of our experience beyond the already familiar and "normalized" modernist techniques of montage, surrealism, and the absurd, for example? In short, rather than trying hard to pursue the anti-sublime ideal, data visualization artists should also not forget that art has the unique license to portray human subjectivity—including its fundamental new dimension of being "immersed in data."

Notes

1.   Of course, if we also think of all 3D computer animation as a type of data visualization in a different sense—for instance, any 3D representation is constructed from a data set describing the polygons of objects in the scene or from mathematical functions describing the surfaces—the role played by data visualization becomes significantly larger. After all, 3D animation is routinely used in industry, science, and popular culture. But I don't think we should accept such an argument, because 3D computer images closely follow traditional Western perspectival techniques of space representation and therefore, from the point of view of their visual appearance, do not constitute a new phenomenon.

2.   Most mappings in both science and art go from nonvisual media to visual media. Is it possible to create mappings that will go into the opposite direction?

3.   Read "against the grain," any descriptive or mapping system that consists of quantitative data—a telephone directory, the trace route of a mail message, and so on—acquires both grotesque and poetic qualities. Conceptual artists explored this well, and data visualization artists may learn from these explorations.

Works Cited

Fry, Benjamin. *Anemone*. 2004. http://acg.media.mit.edu/people/fry/anemone/.

Galloway, Alex. *Carnivore*. 2002. http://rhizome.org/object.php?o=2904&m=2408.

Jevbratt, Lisa. *1.1(2)*. 1999–2002. http://128.111.69.4/~jevbratt/1_to_1/index_ng.html.

———. *Mapping the Web Infome*. 2001. http://128.111.69.4/~jevbratt/lifelike/.

Manovich, Lev. *The Language of New Media*. Cambridge, Mass.: MIT Press, 2001.

Whitney ArtPort. Whitney Museum of American Art. 2002. http://artport.whitney.org/exhibitions/index
    .shtml.

2

Softvideography: Digital Video as Postliterate Practice

Adrian Miles

I open the box to unveil my new home computer. It might be portable, it might not, but if I'm at all interested in making my new purchase the "digital hub" of my new "digital lifestyle," then my computer probably has several USB ports, an IEEE 1394 (also known as FireWire or iLink) port, DVD burner, and if I went for all the options, 802.11b or 802.11g wi-fi and Bluetooth. What this means, outside of the lifestyle advertising that accompanies such hardware, is that it is now technically trivial for me to connect my IEEE 1394–enabled domestic video camera to my computer, capture high-quality, full-resolution video, edit this video, and then print this video back to tape or export it in a digital format for DVD authoring, for e-mail, or to put online. But, aside from digital home movies, what would I now do with all this audiovisual empowerment? In this chapter I'd like to suggest two answers to this question; one looks backward to our existing paradigms of video production, distribution, and presentation, while the other looks, if not forward, then at least sideways to recognize that desktop networked technologies offer novel alternatives not only for production and distribution but also for what constitutes video within networked digital domains. This possible practice treats video as a writerly space where content structures are malleable, variable, and more analogous to hypertext than to what we ordinarily understand digital video to be. I call this practice "softvideography."

Digitization and Production

The influence of digitization on film production is well documented, rampant, and certainly shows no signs of abating (McQuire). These large scale changes in the film and television industries are affecting all sectors of the industry, from big Hollywood features to low-budget independent documentary, yet these changes generally maintain cinema and television as a specific cultural and aesthetic institution, so what has been affected are the means and processes of production but not the form itself. However, the rise of domestic audiovisual technologies—for example, software suites such as Apple's iLife quartet (iDVD, iPhoto, iTunes, and iMovie) or Microsoft's Windows Movie Maker—threatens to do for home video what the original Macintosh achieved for the word with the invention of desktop publishing: the rise of desktop video.

While the rise of digitization has encouraged the distribution of access to a wider range of video tools and has clearly affected the distribution of labor and expertise within various cinematic and televisual industries, these desktop tools have largely concentrated on maintaining film and video as hegemonic aesthetic or material objects. This is what I would like to characterize more specifically as the material hegemony of video and film, and this hegemony is maintained by the manner in which digital video tools support existing paradigms of what a video "object" is. This means that video for software designers, users, and consumers is still conceived of as a linear, time-object that consists principally of an image and a sound track. Even where multiple tracks may be used in what is professionally recognized as post-production—image and sound editing, sound design, effects, and so on—these are generally "burnt" down, much like layers in Adobe's Photoshop, for final delivery.

This hegemony has been maintained in teaching video and cinema, where it is common for vocational and professionally or industry oriented programs to utilize these technologies in the same manner as the broadcast and film industries.

Before exploring and demonstrating some of the potential consequences of this and its implications for teaching in professional, vocational, and creative programs, I'd like to contextualize this paradigm shift using the example of print. This may appear odd, given the evident and obvious distance between print and the moving image; however, I believe that the example of print, digitization, and hypertext has a great deal to teach image-based new media practices. As I've argued elsewhere (Miles, "Cinematic Paradigms"), hypertext can be viewed as a postcinematic writing practice in its combination of minor meaningful units (i.e., shots and nodes) and their possible relations (i.e, edits and links). A great deal of the theoretical issues presented by multilinearity and narrative, whether fiction or nonfiction, including structure, causation, readerly pleasure, closure, repetition, and coherence, have a long and sophisticated history of analysis and work in hypertext. For example, many of the essays in Martin Rieser and Andrea Zapp's anthology on new narrative and cinema, *New Screen Media*, mirrors the excitement and anticipation that hypertext authors and theorists experienced in the early 1980s. Indeed, the isomorphism of the arguments and claims in some of the essays is sadly uncanny, where you could substitute "interactive video" for "hypertext" and not, in fact, notice any difference!

The isomorphic relation that exists between hypertext theory and the new wave of interactive video theory represents a traditional theoretical blindness toward the cognate discipline of hypertext in image-based new media practices and lets us anticipate, on the model of hypertext, the three historical waves of criticism that will happen within interactive video. The first, of which Rieser and Zapp's anthology is a good example, is the work that is produced by those who establish the field and who primarily see a rich series of possibilities and a yet-to-be-invented future. This work is full of excess, anticipation, and an almost naïve expectation about the implications of the technologies for audiences, genres, and media futures. The second wave will largely react against this first wave and will offer critiques of interactive video on the basis that interactivity isn't *really* as interactive as it is scripted, that linearity is inevitable at the micro level in terms of some sort of minimal narrative structural unit, that linearity is an inevitable consequence of readerly actions, and that there have been numerous historical antecedents for the work anyway. Finally, this will mature into a third wave of theory, probably dominated by a second and younger generation of scholars and practitioners, which will accommodate and accept the idealism of the first wave but adopt a much more theoretically pragmatic attitude toward interactive video in relation to its possible media histories and futures.

This history helps us understand contemporary work in interactive video by providing some larger contexts in which it can be inserted. More significantly, it also provides a short circuit by which we can map, or at least point toward, some possible futures simply by recognizing that the minor disciplinary and definitional wars that will occur—What is interactivity, when is interactive video interactive, and is it still cinema?—are important to the development of the field but only in terms of the productive problems it will generate, rather than the hypostatized positions it will produce.

Softcopy Hardvideo

Diane Balestri, in a canonical 1988 essay, characterized the distinction between using a computer for writing and reading in terms of hardcopy and softcopy. Hardcopy is the use of a computer to write but the page, more or less traditionally conceived, is maintained as the publication medium. This is to use all the benefits afforded by desktop publishing and word processing—spelling and grammar checking, nonlinear editing, cutting and pasting, WYSIWYG design, inclusion of graphics, outlining, typographic design, reproducibility, and the various other formal and informal writing practices that have accrued to these word-based technologies. However, hardcopy retains the material hegemony of the page. Content is still presented in primarily linear forms, the dimensions are relatively stable within a document, documents tend to be single objects, and pagination and textual features such as headers, footers, alphabetization, indices, and tables of contents are enforced to manage usability. Readers and writers are largely constructed via the constraints imposed by the medium; for example, closure, temporal coherence, and linear cause and effect are distinguishing features and have been hypostatized as the major formal properties of writing and narrative.

Softcopy, on the other hand, is the use of the computer for writing where the publication format is understood to be the computer screen associated with a modern graphical user interface. This means that content spaces are no longer pages but screens; they can be multiple, variable in size, altered by the user, and that content can now be presented, not only written, in multilinear and multisequential ways. As has been well described by much of the traditional published literature on hypertext (Bolter; Gaggi; Landow; Lanham), the function of the reader tends to change in such environments. The implications of softcopy for the reader have probably been overstated because there are many reading practices that are multilinear and semirandom (television viewing armed with a remote control springs to mind), whereas traditionally the use of a dictionary, encyclopedia, or lifestyle magazine are also good examples of nonlinear or multilinear reading. However, for writers softcopy has much more substantial implications because writing on the basis that your work lacks a definitive beginning and end and may be read in variable sequences and not in its entirety does deeply affect the authority and task of the writer and the status of the text as a particular kind of object. The example that hypertext, hardcopy, and softcopy provide for desktop video is that the relationship between computing as a practice and the discursive objects authored is the same between word processing and hypertext as between desktop video and interactive videography.

The necessity for desktop video software to adopt a hardcopy paradigm is apparent when video material is to be "published" on film or video tape, as both formats basically require an image track and an audio track, though this has some variation across formats. Such media are quintessentially linear and time-based as they physically require the continuous playing of their substrate through a projection or reading apparatus and so ideally support the development of time-based narrative arts. Of course, it is theoretically possible to have only a single still image presented for the duration of a work—for example, Derek Jarman's 1993 feature *Blue*, which consists of a single image held on screen for seventy-nine minutes—but of course in this case an image is recorded and represented twenty-four, twenty-five, or thirty times a second for the duration of the work. The technical necessity of this serialized reading and display requires any digital video editing software to reproduce this so that, once editing is completed, it can be "printed down" and any native digital file structure matches the material demands of video.

In other words, the majority of the tools that are used domestically, professionally, and pedagogically for editing video and sound on the computer adopt a hardcopy, or as I prefer, hardvideo, paradigm. This hardvideo paradigm is evidenced by the way in which all editing systems assume and provide for publication back to tape and so maintain the video equivalent of hardcopy for video on the computer. Hence, these video and audio editing systems, much like word processing, provide numerous advantages and features compared to analogue editing but do not require us as authors or readers to question or rethink any of our assumptions about video as an object. A simple way to illustrate this is simply to think of frame rate and frames per second. Film has traditionally been standardized to twenty-four frames per second during recording and playback (though technically this is varied during recording to achieve things like fast and slow motion and stop-motion animation), whereas video is either twenty-five or thirty frames per second (PAL or NTSC). However, if the computer were to be the publication environment for a digital video work, what would constitute frame rate? Frame rate exists in digital video largely as a legacy of its hardvideo heritage. In the example of Jarman's *Blue*, to edit this film on a digital edit suite and then to "publish" the film, even to the computer rather than to video or film, would require the editing program to "draw" the image twenty-four, twenty-five, or thirty times a second for its seventy-nine minutes. However, a softvideo environment would have no need to do this, simply because the image is static for seventy-nine minutes on a computer screen, so all a softvideo tool would need to do is to draw it once and then simply hold it on screen for seventy-nine minutes. This is how QuickTime works when it is used as an authoring and publishing environment, and this drawing of the frame once and holding it for a specific duration is an example of the difference between hardvideo and softvideo.

This difference may appear to be only a quantitative difference. In the softvideo example, the image track of this seventy-nine-minute movie would literally only be as big as a single blue image at, let's say, 1152 (768 pixels at 72 dots per inch [dpi]). This image, even at very high quality (little or no compression), would be approximately 100 kilobytes (KB) in size, whereas the hardvideo digital equivalent for this image track would be approximately 600 megabytes (MB). However, once we introduce the network into the softvideo paradigm, this difference is size moves from a quantitative to a qualitative change.

Networks

Pedagogically, the distinction between hardcopy and softcopy in relation to text has, in my experience, proved to be a useful analogy for introducing and illustrating the relation of hardvideo to softvideo. Even where students have regarded themselves as primarily image makers, they are deeply immersed in and interpellated by print literacy, and so it provides a familiar context from which to problematize our commonsense notions of what constitutes a possible softvideo practice. However, Balestri's original work pays little regard to the role of the network, and it is obvious that, while the difference between hard and softcopy, and for that matter hard and softvideo, does offer a paradigmatic shift, the introduction of networked technologies and their associated literacies offers a further and dramatic qualitative change.

In this context, the writer-reader of the Web has become the prosumer of the digital hub, combining consumer electronics with desktop technologies to make and view, produce and listen, distribute and download. Clearly, the network is the most fluid and dynamic environment for this to take place in, and it is in the combination produced by desktop video and the network that allows for the rise of a genuinely videographic discourse. This needs to be a practice that accepts the constraints offered by the network as enabling conditions and will become a form of video writing that, like hypertext before it, will produce a hybrid form that no longer looks backward to existing media forms and instead peers forward toward the future genres that it will invent. What prevents this possible future is largely the constraints provided by adopting television or cinema as the primary index defining desktop video as a practice.

Softvideography

Once the computer screen and the network are regarded as the authoring and publication environment for softvideo, video can be treated as hypertext and, in the case of QuickTime, digital video moves from being a publication environment to a writerly environment. This ability to write in and with the multiple types and iterations of tracks that constitute the QuickTime architecture is the basis for softvideography. There is, perhaps, some irony in this because in the past I have argued strongly that hypertext systems, particularly sophisticated stand-alone environments such as Eastgate's Storyspace, are postcinematic writing environments (Miles, "Cinematic Paradigms"). In this postcinematic conception of hypertext, nodes are structurally equivalent to cinematic shots in that they are minimal, meaningful structural units, and links operate much as cinematic edits, joining these meaningful units into larger syntagmatic chains. Of course, the difference in hypertext, unlike traditional cinema or television, is that the syntagmatic chains formed are singular, whereas in hypertext they are plural and their particular formations can vary between the reader, the author, and the system. This is, in fact, a claim I regularly make with my undergraduate students undertaking hypertext work, and after ten years I still have not had a student, all of whom have had some form of audio or video editing experience, not understand that this is, in fact, a very simple but powerful way to consider hypertext content and authoring. However, in softvideography this cinematic hypertextuality is returned to cinema but in a manner that means considerably more than linked movies. To illustrate this I will use the example of hypertext.

As a writer within a hypertext system, it is useful to consider each node as a minimal meaningful unit that may or may not appear in any particular reading. This approach to writing hypertext encourages the writers to shift their conception of authorship and reading away from a print and televisual model, which assumes that the user or reader will comprehensively read or watch the entire work from beginning to end. It also means the writer needs to recognize that each node exists within an economy of individuated readings in which the link structures authored or generated constitute the possibility for this economy. The content of a node has a meaningful status outside of its specific or possible link structure; however, nodes gain their meaning by virtue of the larger structures that they form via links. Simply, each node is a variable that, in conjunction with its links, may or may not appear in an individual reading. Whether they get used or read is subject to various authorial, readerly, and scripted decisions. They may appear or they may not, and when they appear they can be varied. The same consideration of content spaces is necessary in softvideography, so that the softvideograph equivalent of nodes *is not* conceived of as fixed or essential units or blocks but as available and possible. This does require a shift in conception, largely because in such a temporally subjected media as video all shots or other significant units within the work are regarded as essential and inevitable. Shot sixteen is always shot sixteen and because of this is thought of as in some manner quintessentially necessary to the work.

QuickTime is the structural architecture that supports softvideography in this way. In general, QuickTime provides two broad ways of treating content so that a movie's structural units are contingent and variable and not immutable and fixed. The first way is to take advantage of QuickTime's track-based architecture, in which a QuickTime file can consist of not only numerous track types, including video, audio, text, color, sprite (which is a fully programmable track type in QuickTime), midi, and picture, but also multiple instances of each track type. Therefore, it is possible to make a QuickTime movie that contains three video tracks, four soundtracks, a text track, and two picture tracks. Each of these tracks is an object within the movie, and as an object it is structurally identical to how I have characterized a node, so that each track object—for instance, each of the three video tracks—can be made available within the single movie on a contingent or programmatic basis. This means that you can script a QuickTime movie so that each of the three video tracks plays or some combination of the three plays. Similarly, you may play all soundtracks at once or some combination of these, and of course, you can vary the volume of each of the soundtracks subject to system, user, movie, or external variables. This applies to each of the track types and individually for each track, and all of these can be varied in time as the QuickTime file plays, which obviously suggests that complex permutations and relations are possible between all of the tracks of the QuickTime file.

An example of this is "Collins Street" (Miles, "Vog"), which is a small QuickTime work that consists of nine video tracks, three sound tracks, one sprite track, and a color track. The sprite track contains nine still images that are temporarily collaged over individual video panes, and the color track is simply the movie's black background, which in QuickTime is not an image but draws a color at a specified size. As "Collins Street" (a downtown Melbourne street) plays, the user can mouse over each of the video panes and doing so "triggers" the sprite track, which turns on and displays for a prescribed duration a jpeg image that contains text. The same sprite track also controls which of the three simultaneous soundtracks is being heard and its relative volume. While this particular work might be thought of as an experimental documentary, it does illustrate some of the things that can be done using QuickTime and the way in which tracks can be considered "independent" objects within the movie, so that the movie now becomes not a linear audio and visual track but a container for a multiplicity of such tracks that are enabled variably.

As a more complex example, imagine a video image of a student cafeteria with several tables of animated conversation in view. Mousing over each table could, for example, allow the user to hear the conversation at that particular table, while clicking on each table could load a new QuickTime movie that would take you to that table. To make this example more sophisticated, imagine that, within this cafeteria scene, *when* you click on a particular table to zoom in to that specific conversation is significant—to click on a specific table in the second thirty seconds loads a different movie than if you had clicked in the first thirty seconds. Once you begin to appreciate that this is possible, then the sorts of narratives and content that can be made become distinctly different than our existing conceptions of video narrative. Time-dependent or otherwise variable links, embedded within the field of the video, shift authorial activity away from the button model common to multimedia and most existing forms of online video. These contextual intravideo links are qualitatively different sorts of link events than navigational, volume, and play buttons in the same manner that text links within a written hypertext work are qualitatively different than those links constructed and provided by a navigational menu (Ricardo).

The second manner in which QuickTime supports work like this is through its provision of "parent movies" and "child movies." A parent movie is a container movie that may, like any other QuickTime movie, consist of numerous tracks, but it will also include one or more movie tracks. A movie track, which should not be confused with a video track, is a QuickTime track that allows you to load external QuickTime content, in effect other QuickTime movies, into another movie. The movie that contains the movie track is known as the parent movie, and the content that is loaded within the parent movie is known as a child movie. Child movie content can be any data type that QuickTime can read, and it can reside anywhere that the parent movie can access, so if the parent movie is designed to be delivered via the network, then the child movie content can, literally, reside anywhere else on the network. A parent movie can contain multiple child movie tracks, but more impressively, an individual movie track in a parent movie operates as a list so that it may contain numerous individual external files. For example, you can make a QuickTime parent movie that contains a child track, and that individual child track consists of, let's say, a list of nine sound tracks. The parent movie can be scripted so that one of the nine child movies is loaded subject to whatever conditions or actions are scripted for, and this can be altered dynamically during the playing of the parent movie. Child movies exist in complex relations to parent movies because it is possible to tie a child movie's duration and playback to its parent or for the child to be independent of the parent. Where a child movie is slaved to the parent movie, it may only play when the parent movie is playing and it will stop playing when the parent movie ends. Where a child movie track is not slaved, then it can play independently of the parent movie's duration and even separately from the parent movie's play state, so that even where a parent movie may be paused, the child movie can continue to play.

One example of this is "Exquisite Corpse 1.1" (Miles), which is a triptych that has a single parent movie that loads one child movie in each of three movie tracks. Within this brief work the movie tracks appear as video panes in the parent movie, but because they are child movies, the three movies that appear all reside outside of the parent movie. The child movies have been scripted to loop and for their duration to be independent of the parent movie, which in this case is a QuickTime movie that is only one frame long. In addition the bar above and below each video pane is a sprite track, so that mousing into any of the bars controls the volume and the playback rate of each of the three child movies, such that the current video pane plays at twenty-four frames per second at full volume, then the next plays at twelve frames per second at zero volume, and the next at six frames per second also at zero volume. Each of the three movies varies slightly in content, and the effect of this structure means that to view the movie the user literally plays the movie, and that when and where they use the mouse controls the combinations formed between each of the three simultaneous video panes. This has several rather intriguing consequences. The first is that as each of the three child movies have durations that are independent from the parent movie and from each other, then the work as a whole would appear to be endless or at least of indeterminate duration. This is not simply the looping used in animation and programming and described in detail by Lev Manovich because there is no sense of necessary or inevitable repetition involved. The work continues, indefinitely and variably, until the user stops, and while they play it loops, but the combinations formed, the rates of playback, and what is heard and seen will always be novel. The sense of duration implied here is fundamentally different than that associated with film and video, which have traditionally been subject to and by time.

Another implication of this structure is that if we consider montage as the relations between images in time, then here montage is apparent not only within each video but also in the ongoing relations between each video pane, and that this larger set of relations is partially controlled by the user. Hence montage, which is ordinarily conceived of as fixed and immutable, has become unfixed and mutable, which in turn provides a preliminary illustration of how the "place" or "site" of the event of montage will move toward the user in interactive video. This is analogous to the manner in which hypertext theory conceives of the reader's role in relation to the realized text, so that the discursive system becomes a field for the provision of possibilities and individual readings or playings become the realization of individual variations within this field of possibilities.

Softvideo Pedagogy

There are several software packages available at the time of writing that support using QuickTime as an authorial and writerly environment. Some of these tools are cross-platform; however, much of the innovation in interactive video appears to be developing around Apple's OS X operating system, and QuickTime remains the only widespread audiovisual file structure that conceives of time-based media as something other than a delivery platform. QuickTime Pro, which is what QuickTime becomes when you pay the license fee and register your existing copy, provides access to a great deal of these authoring possibilities. EZMediaQTI is a recently developed commercial software package that provides a very simple interface to much of QuickTime's underlying programmable architecture, while Totally Hip's cross platform LiveStage Professional is currently the major tool used in QuickTime authoring, outside of programming QuickTime in Java. However, it is not the software product that is the tool in the context of interactive networked video but QuickTime as an architecture, and considering text on a page as an architecture, the specific tools are less significant than developing literacies around what the architecture makes possible. It is these literacies that not only allow us to use these software products as tools but also let us appropriate them for novel uses and possibilities. After all, one of the major issues confronting teaching technologies in networked and integrated media contexts is the balance and confusion students experience between learning the tool and learning what can be done with the tool. I use three simple exercises to help students move their attention away from the software as an apparatus and toward their "content objects" as the apparatus. Or, to return to my earlier terms, to stop thinking of the software as the architecture and to understand that the architecture is a combination of what and how the work is built.

The first exercise uses QuickTime Pro only and is intended to show that you can make collaged and montaged time-based video works using very simple technologies. The introductory tutorial for this exercise, including all the necessary content, is published online (Miles, "Desktop Vogging") and demonstrates how to import still images into a QuickTime movie, scale the image to a nominated duration (for example, to accompany a soundtrack), embed other still images to create a collage, and embed video tracks over these image tracks to end up with a collaged movie that contains four still images, one soundtrack, and three video tracks. While demonstrating the desktop nature of such work—after all, QuickTime Pro is a simple, thirty-dollar piece of software—it also foregrounds the manner in which tracks in QuickTime are samples and fragments of a larger whole and not completed content that is exported via QuickTime for publication. After this tutorial, students are then invited to collect their own material, using domestic camcorders, digital cameras, scanners, and minidisk, and to make a short, stand-alone QuickTime collage. As their experience builds, constraints are introduced to the exercise, so that the final work must be two minutes in length and may be limited, for example, to a total of 2 MB, containing a nominated number of specified track types. This aspect of the task is where one facet of network literacy is introduced as bandwidth as a meaningful limit becomes concrete under such constraints.

The second exercise is based on this QuickTime collage project that students have already completed and uses QuickTime's HREF track type, which allows a movie to load Web-based content as it plays. The HREF track is a specific type of text track within QuickTime that contains a combination of timecodes and URLs. Specific moments in the film are nominated by timecodes, and the URLs can be either "manual HREFs" or "auto HREFs." For example, this is an extract from an HREF track indicating an in and out point for each HREF the presence of an "A" before a URL indicates that it is an automatic HREF, and these will be loaded automatically as the movie plays. Manual HREFs require the user to click on the video during the relevant interval defined by the timecode in the HREF track.

[00:00:00.00]
http://hypertext.rmit.edu.au/vog/vlog/

[00:00:02.00]
http://www.rmit.edu.au
[00:00:04.00]
A<http://hypertext.rmit.edu.au/vog/>
[00:00:06.00]
A<http://www.rmit.edu.au/adc/mediastudies/>T<newframe>
[00:00:10:00]

The usual way to use a QuickTime movie with an HREF track is to embed the movie on a Web page within a frame and to use the HREF track to target another frame so the URLs contained in the HREF track are automatically displayed in the target frame as the movie plays. The URL that the HREF track indicates is simply a Web page and therefore can contain or display any content that can be displayed via a Web server and within a Web browser, including, of course, other QuickTime movies.

The task for the students is to write a series of text-only Web pages that the Quick-Time file loads as it plays within the frame and for these pages to complement the collaged QuickTime work in a meaningful way. They may, of course, make a new QuickTime collage piece for this task. Text only is nominated because it loads much more quickly than any other sort of content via http, and so it is viable for this project when the work is viewed on the Internet. It is also to encourage students to begin to think about the relation of word to image in terms of aesthetic practice and narrative and, of course, to model the idea that text may exist within the QuickTime movie as embedded, concealed, and operative code rather than surface effect and narrative. This assignment also provides a very minimal form of explicit interactivity between the user, the QuickTime movie, and the loaded pages, particularly where HREF or a combination of HREF and automatic HREF URLs are used, and this requires students to extend their understanding of the possible relations between parts outside of the example of the single QuickTime collage and toward external objects and their possible relations to their movies.

The HREF track is ordinarily written using a simple text editor, imported into QuickTime, converted into an HREF track, exported from QuickTime with a timecode added, and then edited for reimporting into the movie. This is clumsy, and there are more efficient ways of doing this, but it also demystifies what a text track and an HREF track is in a QuickTime movie and insistently demonstrates the desktop nature of softvideography as a writerly practice because, in this example, an interactive Web-based, mixed-media movie has been made using only QuickTime Pro and whatever free text editor comes with the computer's operating system.

The third exercise is also network-based and is created to help students think about and understand the possible relations between parts and the implications of this. While the intention of softvideography is to use QuickTime as a writerly environment, this does extend beyond the internal relations of tracks and samples to include the relations of external tracks or samples, which are, of course, important when working with parent and child movies, as well as an understanding of multilinear environments in general. The formal task, which can involve video, still images, or a combination of both, is to develop a narrative that consists of seven shots in which each of the seven shots may appear at any point in the sequence. In other words each shot or image may appear once at any point in the sequence of seven and, regardless of where it appears, the sequence must still retain narrative coherence. Intertitles can be used, though this counts as a shot. The students then embed their video on a Web page that contains a script that automatically randomizes the insertion of each of the seven movie files. This is done by taking advantage of QuickTime's QTNext tag, available when QuickTime is embedded via HTML, which allows you to play up to 256 QuickTime files in a row, so that as the first QuickTime file ends, the QuickTime plug in then requests the next file, and so on. This means that when the Web page that contains their movies is viewed, each individual viewing displays one of 5,040 possible sequences.

This exercise is useful because it allows students to see how complex narrative or multilinear possibilities develop from quite simple and small sets and that complexity is not synonymous with the large scale nested or branching structures that is common when students first start trying to conceive of multilinearity. This task also demonstrates one of the most difficult aspects of new media narration because it is a particularly demanding task to conceive of seven possible moments, images, shots, or words that can be randomly presented in this manner and yet retain something that might be identified as narrative. Many of the works produced are more like tone poems or mood pieces, what film semiotician Christian Metz has catalogued as a bracket syntagma, which, of course, suggests the difficulty of constituting narrative via fragments that need to be narratively "permeable" in this manner. Incidentally, this exercise also helps students in their reading of canonical hypertext literature, such as Michael Joyce's *Afternoon* or Stuart Moulthrop's *Victory Garden*, because it provides them with a cognitive and formal template to understand the structural problems and processes of these works. This might indicate that narrative is not a reasonable expectation of work that is intended to be as multivalent as this.

Conclusion

When we learn to treat desktop digital video as a writerly space, with all the baggage that this connotes, we can recognize that an architecture such as QuickTime is a programmatic environment and, like many other networked programmatic environments, involves consideration of the relations between image and word. This is the crux of what constitutes new media and networked literacy and is why digital video as a networked, distributed, and writerly practice becomes an exemplar for the issues confronting teaching and learning in contemporary media contexts. Softvideography reconfigures the relation of author to viewer in the domain of time-based media and provides one model for a future pedagogy that emerges from the implications of networked digital practice. Such tools not only allow us to reconsider what could be done with video and sound online but also offer the possibility for developing novel expressions of learning and knowledge. This is an ambitious agenda but one that our students deserve and require for the networked ecology that they are inheriting.

Works Cited

Balestri, Diane Pelkus. "Softcopy and Hard: Wordprocessing and the Writing Process." *Academic Computing* Feb. 1988: 41–45.

Bolter, Jay David. *Writing Space: The Computer, Hypertext, and the History of Writing*. Hillsdale, N.J.: Lawrence Erlbaum Associates, 1991.

Gaggi, Silvio. "Hyperrealities and Hypertexts." *From Text to Hypertext: Decentering the Subject in Fiction, Film, the Visual Arts, and Electronic Media*. Philadelphia: University of Pennsylvania Press, 1997. 98–139.

Joyce, Michael. *Afternoon: A Story*. Watertown, Mass.: Eastgate Systems, 1987.

Landow, George P. *Hypertext 2.0: The Convergence of Contemporary Critical Theory and Technology*. Baltimore: John Hopkins University Press, 1997.

Lanham, Richard A. "The Electronic Word: Literary Study and the Digital Revolution." *The Electronic Word: Democracy, Technology, and the Arts*. Chicago: University of Chicago Press, 1993. 3–28.

Manovich, Lev. *The Language of New Media*. Cambridge, Mass.: MIT Press, 2001.

McQuire, Scott. *Crossing the Digital Threshold*. Brisbane: Australian Key Centre for Cultural and Media Policy, 1997.

Metz, Christian. *Film Language: A Semiotics of the Cinema*. Trans. Michael Taylor. New York: Oxford University Press, 1974.

Miles, Adrian. "Cinematic Paradigms for Hypertext." *Continuum: Journal of Media and Cultural Studies* 13.2 (July 1999): 217–26.

———. "Desktop Vogging: Part One." *Fine Art Forum* 17.3 (2003). http://www.fineartforum.org/Backissues/Vol_17/faf_v17_n03/reviews/desktopvogging/index.html.

———. "Exquisite Corpse 1.1." *Videoblog: Vog* 2002. 19 Feb. 2003. http://vogmae.net.au/content/view/56/28/.

———. "Vog: Collins Street." *BeeHive* 4.2 (2001). http://beehive.temporalimage.com/archive/42arc.html

Moulthrop, Stuart. *Victory Garden*. Watertown, Mass.: Eastgate Systems, 1991.

Ricardo, Francisco J. "Stalking the Paratext: Speculations on Hypertext Links as a Second Order Text." *Proceedings of the Ninth Acm Conference on Hypertext and Hypermedia: Links, Objects Time and Space—Structure in Hypermedia Systems*. Eds. Frank Shipman, Elli Mylonas, and Kaj Groenback. Pittsburgh: ACM, 1998. 142–51.

Rieser, Martin, and Andrea Zapp, eds. *New Screen Media: Cinema/Art/Narrative*. London: British Film Institute, 2002.

3

Technopolitics, Blogs, and Emergent Media Ecologies: A Critical/Reconstructive Approach

Richard Kahn and Douglas Kellner

Since the blossoming of hypertext and the the Internet from the early 1990s, the emergence of a utopian rhetoric of cyberdemocracy and personal liberation has accompanied the growth of the new online communities that formed the nascent World Wide Web. While the initial cyberoptimism of many ideologues and theorists of the "virtual community" (Rheingold, "The Virtual Community"; Barlow; Gates; Kelly) now seems partisan and dated, debates continue to rage over the nature, effects, and possibilities of the Internet and technopolitics.[1] Some claim that the Internet's role, as the primary engine driving the ecological arrangement of today's new media, is simply to produce a proliferation and cyberbalkanization of "daily me" news feeds and fragmented communities (Sunstein; Van Alstyne and Brynjolfsson), while others argue that Internet content is often reduced to the amplification of cultural noise and effectless content in what might be termed a new stage of "communicative capitalism" (Dean, "Communicative Capitalism").[2]

In our view, the continued growth of the Internet and emergent media ecologies ultimately have to be thought through together as a complex set of digital tools for organizing novel relations of information and global-local cultural interaction. Contemporary "media ecologies" extend Marshall McLuhan's notion of media as environments that constantly evolve as new media and technologies appear (McLuhan, *Understanding Media*). While people have lived in natural place-based ecologies for a long time and have mixed elements of place with those of industrialized cultural space in modern urban ecologies, today's media ecologies relate people throughout the globe and constitute a virtual world space that is a complex amalgam of ever-shifting global and local spaces and places. In this networked and interconnected world, emergent media ecologies exert and contain a variety of sociopolitical, cultural, and historical forces that interact with people as they become media producers and consumers. In our view, this constantly evolving and mutating media ecology compels people to understand, negotiate, struggle with, and ultimately transform contemporary technology and society.

If emergent media are to remain tools for human uses rather than instruments of mass dehumanization, then the technopolitics of such emergent media ecologies must be continually retheorized from a standpoint that is both critical and reconstructive and be subject to active transformative practice. This approach should be critical of corporate and mainstream forms and uses of technology and should advocate for the reconstruction, or redeployment, of such technologies to further the projects of progressive social and political struggles. Yet, we do not mean to imply that the technopolitics we engage here are essentially, or even mostly, participatory and democratic. We recognize major commercial interests at play and that emergent technologies are presently the site of a struggle involving competing groups from the Far Right to the Far Left on the political spectrum. Further, it is clear that as media and technology evolve in such a way that technoculture and culture come to be more closely one and the same, decisive political issues require answers about the role of public participation in Internet design and access to how individuals and groups will use and configure information and communication technologies (Feenberg, *Alternative Modernity*; Luke, "Cyber-schooling and Technological Change"; Winner). Recognizing the many ways in which politics becomes limited as it implodes into technoculture, we therefore want to engage in a dialectical critique of how emergent types of information and communication technologies (ICTs) have facilitated oppositional cultural and political movements and provided possibilities for the sort of progressive social change and struggle that is an important dimension of contemporary cultural politics.

Alternative Globalizations: Global/Local Technopolitics

Sociologically, the "Information Society" signifies a dynamic and complex space in which people can construct and experiment with identity, culture, and social practices (Poster; Turkle). It also represents the possibility of cultural forms like the Internet to make more information available to a greater number of people, more easily, and from a wider array of sources than any instruments of information and communication in history (Kellner, *Media Spectacle and the Crisis of Democracy*). On the other hand, information and communication technologies have been shown to retard face-to-face relationships (Nie and Ebring), threaten traditional conceptions of the commons (Bowers), and extend structures of Western imperialism and advanced capitalism to the ends of the earth (Trend). Right-wing and reactionary forces can and have used the Internet to promote their political agendas as well. Indeed, one can easily access an exotic witch's brew of Web sites maintained by the Ku Klux Klan and myriad neo-Nazi assemblages, including the Aryan Nation and various militia groups. Internet discussion lists also disperse these views, and right-wing extremists are aggressively active on many computer forums.[3]

These organizations are hardly harmless, having carried out violence of various sorts extending from church burnings to the bombings of public buildings. Adopting quasi-Leninist discourse and tactics for ultra-Right causes, these groups have been successful in recruiting working-class members devastated by the developments of global capitalism, which has resulted in widespread unemployment for traditional forms of industrial, agricultural, and unskilled labor. Moreover, extremist Web sites have influenced alienated middle-class youth as well. The 1999 HBO documentary "Hate on the Internet" provides a disturbing number of examples of how extremist Web sites influenced disaffected youth to commit hate crimes and extremist Islamists have regularly used the Internet to promote anger and Jihad (Jordan, Torres, and Horsburgh). In fact, as Web sites like http://www.alneda.com attest, a particularly disturbing twist in the saga of technopolitics seems to be that global "terrorist" groups are now increasingly using the Internet and Web sites to document, promote, and coordinate their causes (Kellner, *From 9/11 to Terror War*).

Alternatively, as early as 1986, when French students coordinated a national strike over the Internet-like Minitel system, there are numerous examples of people redeploying information technology for their own political ends, thereby actualizing the potential for a more participatory society and oppositional forms of social organization (Feenberg, *Alternative Modernity* and *Questioning Technology*). Since the mid-1990s, following the Zapatista Movement's deployment of the Internet to enlist global support for their regional struggle, there have been growing discussions of Internet-related activism and how information and communication technologies have been used effectively by a variety of progressive political movements (Best and Kellner, *The Postmodern Adventure*; Meikle; Couldry and Curran; Jenkins and Thorburn; McCaughey and Ayers). Infamously, in the late 1990s, activists began employing the Internet to foster movements against the excesses of corporate capitalism. A global protest movement surfaced "from below" (Brecher, Costello, and Smith; Steger) in resistance to the World Trade Organization (WTO) and related globalization policies, while championing democratization and social justice, and this movement has resulted in an ongoing series of broad-based, populist political spectacles such as the "Battle For Seattle" and the unprecedented public demonstration of millions around the world on February 15, 2003, demanding peace and an end to war.

Recent advances in personal, mobile informational technology, combined with widely syndicatable new HTML forms like blogs, are rapidly providing the structural elements for the existence of fresh kinds of global and local technoculture and politics (Rheingold, *Smart Mobs*). As Howard Rheingold notes, the resulting complex of multiuser networks has the potential power to transform the "dumb mobs" of totalitarian and polyarchical states into "smart mobs" of socially active personages linked by notebook computers, personal digital assistant (PDA) devices, Internet cell phones, pagers, and global positioning systems (GPS). New media ecologies such as these provide an important challenge for developing a critical theory of globalization and its contestations.

From the perspective of contemporary technopolitics, it can be argued that dichotomies such as the global and the local express contradictions and tensions between crucial constitutive forces of the present age; therefore, it is a mistake to reject a focus on one side in favor of an exclusive concern with the other (Cvetkovich and Kellner; Kellner, "Globalization and the Postmodern Turn"; Castells; Kellner, "Theorizing Globalization"). Hence, emergent media ecologies force us to think through the relationships between the global and the local by specifically observing how global forces influence and structure an increasing number of local situations and how local forces inflect global forces to diverse ends and produce unique configurations of the global-local as the matrix for thought and action in media culture (Luke and Luke, "A Situated Perspective on Cultural Globalization").

In an important example, mobile global-local digital media were essential to Spain's March 2004 mobilizations wherein activists utilized them to quickly organize people for a mass demonstration against the conservative government's official account of (and indirect role in) the Madrid subway terrorist attack. With haste, mobile technologies such as cell phones, e-mail, text-messaging, and Web sites were also used to achieve a stunning political upset when the antiwar Socialist Party was elected into office because of a massive online and offline "get out the vote" campaign. It is our contention here that examples such as this are becoming increasingly common and the intersections between the ICTs, other digital tools like cameras, mass populaces, and democratic politics represent powerful new networked spaces for the progressive reconstruction of social life. Thus, while emergent media tools like Internet-ready PDAs can provide yet another impetus toward experimental identity construction and identity politics when let loose in a technoculture, the multifaceted social ecologies that they constitute also link diverse communities such as labor, feminist, ecological, peace, and various anticapitalist groups, thereby providing the basis for a broadly democratic politics of alliance and solidarity to overcome the limitations of postmodern identity politics (Dyer-Witheford, *Cyber-Marx*; Best and Kellner, *The Postmodern Adventure*; Burbach, *Globalization and Postmodern Politics*).

Another telling example of how new digital technologies of everyday life are transforming contemporary U.S. politics comes from the role of YouTube in the 2006 U.S. Congressional elections. Senator George Allen (R-Va) was heavily favored to win reelection to the Senate and was being touted as a 2008 Republican presidential candidate when he fell afoul of the ubiquity of digital media on the campaign trail. Allen was caught on tape baiting a young man of color, who was doing oppositional video, and calling him "macaca." The video was put on YouTube and then sent through the Internet, eventually emerging on network television. The YouTube footage became part of a spectacle that coded Allen as a racist and he eventually lost his reelection bid.

Contemporary Internet-related activism is arguably an increasingly important domain of current political struggles that is creating the base and the basis for an unprecedented worldwide antiwar, pro-peace, and social justice movement during a time of terrorism, war, and intense cultural contestation. Examples of oppositional use of emergent technologies have been regularly occurring with the anticorporate globalization, antiwar, and progressive social movements, and all of these together demonstrate that, while it is significant to criticize the ways in which emergent media ecologies can serve as one-dimensionalizing environments, it is equally necessary to examine the ways in which everyday people subvert the intended uses of these media tools (and so those who produce them) for their own needs and uses.[4] Correspondingly, new media such as the Internet have undergone significant transformations during this time toward becoming more participatory and democratic themselves. Innovative forms of communicative Web-based design, such as blogs, wikis, and social networking portals, have emerged as central developments of this trend.[5]

Blogs, Wikis, and Social Networking: Toward Alternative and Participatory Global-Local Interventions

Emergent interactive forms of technopolitics, such as blogs and wikis have become widely popular Internet tools alongside the ultimate "killer app" of e-mail. Wikipedia and other wikis that we describe have become major sites of collective information production and dissemination. The mushrooming community that has erupted around blogging is particularly deserving of analysis here, as bloggers have repeatedly demonstrated themselves as technoactivists favoring not only democratic self-expression and networking but also global media critique and journalistic sociopolitical intervention.

Blogs, short for "Web logs," are partly successful because they are relatively easy to create and maintain—even for Web users who lack technical expertise. Combining the hypertext of Web pages, the multiuser discussion of message boards and listservs, and the mass-syndication ability of Really Simple Syndication (RSS) and Atom Syndication Format (ASF) platforms (as well as e-mail), blogs are popular because they represent the next evolution of a Web-based experience that is connecting a range of new media. The World Wide Web managed to form a global network of interlocking, informative Web sites, and blogs have made the idea of a dynamic network of ongoing debate, dialogue, and commentary come alive both online and offline, emphasizing the interpretation and dissemination of alternative information to a heightened degree.

While the initial mainstream coverage of blogs tended to portray them as narcissistic domains for one's own individual opinion, many group blogs exist, such as Daily Kos (http://www.dailykos.com), Think Progress (http://www.thinkprogress.org), and BoingBoing (http://www.boingboing.net), in which teams of contributors post and comment upon stories. The ever-expanding series of international Indymedia (http://www.indymedia.org) sites, erected by activists for the public domain to inform one another both locally and globally, are especially promising. But even for the many millions of purely individual blogs, connecting up with groups of fellow blog readers and publishers is the netiquette norm, and blog posts inherently tend to reference (and link) to online affinity groups and peers to an impressive degree.

A controversial *New York Times* article by Katie Hafner cited a Jupiter Research estimate that only 4 percent of online users read blogs; however, bloggers were quick to cite a PEW study that claimed 11 percent of Internet users read blogs regularly. Although Hafner's article was itself largely dismissive, it documented the passionate expansion of blogging among Internet users, and the voluminous and militant blogger response to the article showed the blogosphere contained a large number of committed participants. Technorati (http://www.technorati.com), the primary search engine for blogs, claims to track some 28.9 million blog sites, and Technorati's media contact, Derek Gordon, explains that according to their statistics there are 70,000 new blogs per day, 700,000 new posts daily, and 29,100 new posts per hour on average.

One result of bloggers' fascination with networks of links has been the subcultural phenomenon known as "Google bombing." Documented in early 2002, it was revealed that the popular search engine Google had a special affinity for blogs because of its tendency to favor highly linked, recently updated Web content in its site ranking system. With this in mind, bloggers began campaigns to get large numbers of fellow bloggers to post links to specific postings designed to include the desirable keywords that Google users might normally search. A successful Google bomb, then, would rocket the initial blog that began the campaign up Google's rankings to number one for each and every one of those keywords—whether the blogs had important substantive material in them or not.

While those in the blog culture often abused this trick for personal gain (to get their own name and blog placed at the top of Google's most popular search terms), many in the blog subculture began using the Google bomb as a tool for political subversion. Known as a "justice bomb," this use of blogs served to link a particularly distasteful corporation or entity to a series of key words that either spoof or criticize the entity. Hence, thanks to a justice bomb, Google users typing in "McDonald's" might very well get pointed to a much-linked blog post titled "McDonald's Lies about Their Fries" as the top entry. Another group carried out a campaign to link Bush to "miserable failure" so that when one typed this phrase into Google, one was directed to George W. Bush's official presidential Web site. While Google continues to favor blogs in its rankings, amid the controversy surrounding the so-called clogging of search engine results by blogs, it has recently taken steps to de-emphasize blogs in its rating system and may soon move blogs to their own search subsection altogether—this despite blogs accounting for only an estimated 0.03 percent of Google's indexed Web content (Orlowski).

Whether or not they are highly ranked in Google, many blogs are increasingly political in the scope of their commentary. Over the last year, a plethora of leftist-oriented blogs have been created and organized in networks of interlinking solidarity to contest the more conservative and moderate blog opinions of mainstream media favorites like Glenn Reynolds (http://www.instapundit.com). Post-September 11, with the wars in Afghanistan and Iraq, the phenomenon of "Warblogging" became an important and noted genre in its own right. Blogs, such as our own BlogLeft (http://www.gseis.ucla.edu/courses/ed253a/blogger.php), have provided a broad range of critical alternative views concerning the objectives of the Bush administration, the Pentagon, and the corporate media spin surrounding them. One blogger, the now famous Iraqi Salam Pax (http://www.dear_raed.blogspot.com), gave outsiders a dose of the larger unexpurgated reality as the bombs exploded overhead in Baghdad. Meanwhile, in Iran, journalist Sina Mottallebi became the first blogger to be jailed for "undermining national security through cultural activities."[6] And after the 2004 election in Iran, boycotted by significant groups of reformers after government repression, dozens of new Web sites popped up to circulate news and organize political opposition.

In response to the need for anonymous and untraceable blogging, open source software like invisiblog (http://www.invisiblog.com) has been developed to protect online citizens' and journalists' identities. Untraceable blogging is crucial in countries where freedom of speech is in doubt, such as China, which has forced bloggers to register with the government and has jailed journalists, bloggers, and individuals who post articles critical of the government. Recent news that the FBI has begun actively monitoring blogs in order to gain information on citizens suggests a need for U.S. activist-bloggers to implement the software themselves, just as many use PGP (Pretty Good Privacy) code keys for their e-mail and anonymity cloaking services for their Web surfing (http://www.anonymizer.com) and (http://tor.eff.org).

On another note, political bloggers have played a significant role in several significant media spectacles in U.S. politics, beginning in 2003 with the focus of attention upon the racist remarks made by Speaker of the House Trent Lott and then the creation of a media uproar over the dishonest reporting exposed at the *New York Times* in the Jayson Blair scandal. Lott's remarks had been buried in the back of the *Washington Post* until communities of bloggers began publicizing them, generating public and media interest that then led to his removal. In the *New York Times* example, bloggers again rabidly set upon the newsprint giant, whipping up so much controversy and hostile journalistic opinion that the *Times*'s executive and managing editors were forced to resign in disgrace. Likewise, CBS News and its anchor Dan Rather were targeted when right-wing bloggers attacked and debunked a September 2004 report by Rather on *60 Minutes* that purported to reveal documents suggesting that the young George W. Bush disobeyed an order in failing to report for a physical examination and that Bush family friends were enabling him to get out of completing his National Guard service (Kellner, *Media Spectacle and the Crisis of Democracy*).

Another major blog intervention involves the campaign against Diebold computerized voting machines. While the mainstream media neglected this story, bloggers constantly discussed how Republican activists ran the company, how the machines were unreliable and could be easily hacked, and how paper ballots were necessary to guarantee a fair election. After the widespread failure of the machines in 2003 elections and a wave of blog discussion, the mainstream media finally picked up on the story and the state of California cancelled their contract with the company—although Arnold Schwarzenegger's secretary of state reinstated the Diebold contract in February 2006, producing another round of controversy. Of course, this was not enough to prevent Diebold's playing a major role in the 2004 presidential election, and bloggers were primarily responsible for challenging the e-vote machine failures and analyzing alleged vote corruption in states like Ohio, Florida, and North Carolina (Kellner, *Media Spectacle and the Crisis of Democracy*).

Taking note of blogs' ability to organize and proliferate groups around issues, the campaign for Howard Dean became an early blog adopter (http://www.blogforamerica.com), and his blog undoubtedly helped to successfully catalyze his grassroots campaign (as well as the burgeoning antiwar movement in the United States). In turn, blogs became *de rigueur* for all political candidates and have been sites for discussing the policies and platforms of various candidates, interfacing with local and national support offices, and in some cases speaking directly to the presidential hopefuls themselves.[7] Moreover, Internet-based political projects such as http://www.moveon.org, which began during the Clinton impeachment scandal, began raising money and issues using computer technologies in support of more progressive candidates and assembling a highly impressive e-mailing list and groups of loyal supporters.

Another momentous media spectacle, fueled by intense blog discussion, emerged in May 2004 with the television and Internet circulation of a panorama of images of the United States' abuse of Iraqi prisoners and the quest to pin responsibility on the soldiers and higher U.S. military and political authorities. Evoking universal disgust and repugnance, the images of young American soldiers humiliating Iraqis circulated with satellite-driven speed through broadcasting channels and print media, but it was the manner in which they were proliferated and archived on blogs that may make them stand as some of the most influential images of all time. Thanks in part to blogs and the Internet, this scandal will persist in a way that much mainstream media spectacle often does not, and the episode suggests that blogs made an important intervention into Bush's policy and future U.S. military policy and may play an important role in future U.S. politics.

Bloggers should not be judged, however, simply by their ability to generate political and media spectacle. As alluded to earlier, bloggers are cumulatively expanding the notion of what the Internet and cyberculture are and how they can be used. In the process they are questioning, and helping to redefine, conventional journalism, its frames, and its limitations. Thus, a genre of "Watchblogs" (http://www.watchblog.com) has emerged that focuses upon specific news media, or even reporters, dissecting their every inflection, uncovering their spin, and attacking their errors. Many believe that a young and inexperienced White House press corps was hypercritical of Al Gore in the 2000 election, while basically giving George W. Bush a pass, a point made repeatedly by Internet media critic Bob Somerby at http://www.dailyhowler.com/ (see also Kellner, *Grand Theft 2000*).

Since the 2004 election, however, the major media political correspondents have been minutely dissected for their biases, omissions, and slants. One astonishing case brought by the watchblogging community was that the Bush administration provided press credentials to a fake journalist who worked for a certain Talon News service that was a blatant front for conservative propaganda. The Bush White House issued a press pass to avowed conservative partisan "Jeff Gannon," who was a regular in the White House Briefing Room, where he was frequently called upon by Bush administration press secretary Scott McClellan whenever the questions from the press corps got too hot for comfort. After he manufactured quotes by Senators Clinton and Reid in White House press conferences, bloggers found out that Gannon's real name was James Guckert and that he also ran gay porn sites and worked as a gay escort. As another example of the collapse of the investigative functions of the mainstream media, although "Gannon" was a frequent presence, lobbing softball questions in the White House briefing room, his press colleagues never questioned his credentials, instead leaving to bloggers the investigative reporting that the mainstream media was apparently too lazy and incompetent to do themselves (Kellner, *Media Spectacle and the Crisis of Democracy*).

One result of the 2004 election and subsequent U.S. politics has been the decentering and marginalizing of the importance of the corporate media punditocracy by Internet and blogosphere sources. A number of Web sites and blogs have been dedicated to deconstructing mainstream corporate journalism, taking apart everyone from the right-wing spinners on Fox to reporters for the *New York Times*. An ever-proliferating number of Web sites have been attacking mainstream pundits, media institutions, and misreporting; with the possible exception of the *New York Times*'s Paul Krugman, Internet and blog sources were often much more interesting, insightful, and perhaps even influential than the overpaid, underinformed, and often incompetent mainstream corporate media figures. For example, every day the incomparable Bob Somerby on http://www.dailyhowler.com savages mainstream media figures, disclosing their ignorance, bias, and incompetence, while a wide range of other Web sites and blogs contain media critique and alternative information and views (for instance, mediamatters.org).

As a response, there have been fierce critiques of the blogosphere by mainstream media pundits and sources, although many in the corporate mainstream are developing blogs, appropriating the genre for themselves. Yet mainstream corporate broadcasting media, and especially television, continue to exert major political influence, and constant critique of corporate media should be linked with efforts at media reform and developing alternatives, as activists have created media reform movements and continue to create ever-better critical and oppositional media linked to ever-expanding progressive movements. Without adequate information, intelligent debate, criticism of the established institutions and parties, support of the media reform movement, and cultivation of mean-ingful alternatives, democracy is but an ideological phantom, without life or substance.[8]

Democracy requires action, even the activity of computer terminals. However, part of the excitement of blogging is that it has liberated producers and designers from their desk-tops. Far from writing in virtual alienation from the world, today's bloggers are posting pictures, text, audio, and video on the fly from PDA devices and cell phones as part of a movement of "mobloggers" (i.e., mobile bloggers; see http://www.mobloggers.com). Large political events, such as the World Summit for Sustainable Development, the World Social Forum, and the G8 forums, all now have wireless bloggers providing real-time alternative coverage, and a new genre of "confblogs" (i.e., conference blogs) has emerged as a result.[9] One environmental activist, a tree-sitter named Remedy, even broadcast a wireless account of her battle against the Pacific Lumber Company from her blog (http://www.contrast.org/treesit), 130 feet atop an old-growth redwood. She has since been forcefully removed but continues blogging in defense of a sustainable world in which new technologies can coexist with the wilderness and other species.

In fact, there are increasingly all manner of blogging communities. "Milbloggers" (i.e., military bloggers) provide detailed commentary on the action of the United States and other troops throughout the world, sometimes providing critical commentary that eludes mainstream media. And in a more cultural turn, blog types are emerging that are less tex-tual, supported by audio bloggers, video bloggers, and photo bloggers, with the three often meshing as an on-the-fly multimedia experience. Blogging has also become important within education circles (e.g., http://www.ebn.weblogger.com), and people are forming university blogging networks (e.g., http://blogs.law.harvard.edu) just as they previously created citywide blogging portals (e.g., http://www.nycbloggers.com).

While the overt participatory politics of bloggers, as well as their sheer numbers, makes the exciting new media tool called the "wiki" secondary to this discussion, the inherent participatory, collective, and democratic design of wikis have many people believ-ing that they represent the coming evolution of the hypertextual Web. Taken from the Hawaiian word for "quick," wikis are popular, innovative forms of group databases and hypertextual archives that work on the principle of open editing, meaning that any online user can change not only the content of the database (i.e., add, edit, or delete) but also its organization (i.e., the way in which material links together and networks). Wikis have been coded in such a way that they come with a built-in failsafe that automatically saves and logs each previous version of the archive. This makes them highly flexible because users are then free to transform the archive as they see fit, as no version of the previous information is ever lost beyond recall. The result, then, is not only an information-rich data bank but also one that can be examined *in process*, with viewers able to trace and investigate how the archive has grown over time, which users have made changes, and what exactly they have contributed.

Although initially conceived as a simple, informal, and free-form alternative to more highly structured and complex groupware products such as IBM's Lotus Notes, wikis can be used for a variety of purposes beyond organizational planning (Leuf and Cunningham). To the degree that wikis could easily come to supplant the basic model of the Web site, which is designed privately, placed online, and then is mostly a static experience beyond following preprogrammed links, wikis deserve investigation by technology theorists as the next wave in the emerging democratic Web-based media.

One interesting wiki project is the dKosopedia (http://www.dkosopedia.com), which is providing a valuable cultural resource and learning environment through its synthesis and analysis of the connections behind today's political happenings. Perhaps the preeminent example of wiki power, though, is the impressive Wikipedia (http://www .wikipedia.org), a free, globally collaborative encyclopedia project based on wiki protocol that would have made Denis Diderot and his fellow *philosophes* proud. Beginning on January 15, 2001, Wikipedia has quickly grown to include approximately one million always-evolving articles in English (with nearly four million in more than one hundred languages total), and the database grows with each passing day. With over thirteen thousand vigilant contributors worldwide creating, updating, and deleting information in the archive daily, the charge against wikis is that such unmoderated and asynchronous archives must descend into informative chaos. However, as required by the growth of the project, so-called Wikipedians have gathered together and developed their own loose norms regarding what constitutes helpful and contributive actions on the site. Disagreements, which do occur, are settled online by Wikipedians as a whole in what resembles a form of virtualized Athenian democracy wherein all contributors have both a voice and vote.

Further, when it was revealed that corporations and interest groups were intervening and erasing episodes and information from Wikipedia that presented them in a negative light, ways were discovered to find the e-mail addresses of those changing the entries and a Web site Wikiscanner was developed by Virgil Griffith exposing the activities of those who would purge information questioning their activities. As Annalee Newitz describes it:

> Turns out that all the anonymous propaganda and lies on *Wikipedia* aren't coming from basement dwellers at all—they're coming from Congress, the CIA, The *New York Times*, The *Washington Post*, and the American Civil Liberties Union. Somebody at Halliburton deleted key information from an entry on war crimes; Diebold, an electronic-voting machine manufacturer, deleted sections of its entry about a lawsuit filed against it. Someone at Pepsi deleted information about health problems caused by the soft drink. Somebody at The *New York Times* deleted huge chunks of information from the entry on the *Wall Street Journal*. And of course, the CIA has been editing the entry on the Iraq war.

There are other technological developments that are contributing to more bottom-up, participatory, and democratic cultural forms, and among the multiple efforts is the growing ubiquity of digital cameras, which allow instantaneous recording of events and disseminating the images or scenes globally through the Internet. The Abu Ghraib prison torture scandal was documented by soldiers who sent digital pictures to others, one of whom collected and released the incriminating images to the media.[10] Digital cameras and Internet connections thus enable individuals to collect and transmit images and scenes that could undermine existing authorities like the police, the military, corporations, or the state when they are participating in abusive activities. The widespread use of digital cameras and cell phone images and footage was documented in the 2005 Asian Tsunami, the London terrorist bombing, and Hurricane Katrina when early images of the disasters were disseminated quickly throughout the globe on digital cell phones. In 2006, several instances of police brutality were captured on digital cameras and cell phones, leading to mainstream media exposure and disciplining of the guilty parties.

These developments in digital technology further democratize journalism, making everyone with digital technology and Internet connections a potential journalist, commentator, and critic of existing media and politics. Since digital imagery and texts can be instantly uploaded and circulated globally via sites like Indymedia and the wide array of photoblogs, the dissemination of alternative information and views could intensify in the future and continue to undermine the power of corporate media and the state. Likewise, podcasting brings new multimedia forms into the emergent media ecology and helps disseminate self-produced alternative information and entertainment, as well as mainstream content, in the form of audio and video that is searchable and downloadable based on user preferences via the Internet.

Technological developments may encourage more social and political forms. Blogs like Corante's Many2Many (http://www.corante.com/many) track how blogs and wikis are pointing toward a greater trend in emergent media development: "social software" that networks people with similar interests and other semantic connections. As alluded to earlier, Howard Dean's campaign use of Web-facilitated "meet-ups" generated novel forms for grassroots electoral politics enthusiasm, but notably, people are using online social networking to gather based on all kinds of topics and issues (http://www.meetup.com). More recently, social software has moved to incorporate a quasi "six degrees of separation" model into its mix, with portals like MySpace (http://www.myspace.com), Friendster (http://www.friendster.com), Facebook (http://www.facebook.com), LinkedIn (http://www.linkedin.com), Ryze (http://www.ryze.com), Orkut (http://www.orkut.com), and FriendFan (http://www.friendfan.com) allowing groups to form around common interests, while also creating linkages and testimonials between friends and family members. This has allowed for a greater amount of trust in actually allowing virtual relationships to flourish offline while also allowing a new friendship to quickly expand into the preexisting communities of interest and caring that each user brings to the site.

Though all of these examples are reason to hope that the emergent media ecology can be tools for the strengthening of community and democracy among its users, it must be stressed again that we do not conclude that blogs or wikis or social networking software, alone or altogether, are congruent with strong democratic practices and emancipatory anticapitalist politics. For all the interesting developments we are chronicling here, there are also shopping blogs, behind-the-firewall corporate wikis, and all-in-one business platforms such as Microsoft's planned Wallop application. It remains a problem that most blogs, while providing the possibility for public voice for most citizens, are unable to be found by most users, thus resulting in so-called nanoaudiences. Further, the fact that a great many of the millions of blogs have an extremely high turnover rate, falling into silence as quickly as they rise into voice, and that huge amounts of users remain captivated by the individualistic diary form of the "daily me" means that the logic of bourgeois individualism and capitalism is here too apparent.

Although we have pointed to positive effects of blogs, wikis, and new digital technology in the emergent media ecology of the contemporary era, we are aware of downsides and do not want to sacrifice the edge of critique in advocating our reconstructive approach. As blogs proliferate, they also fragment, and a new hierarchy of bloggers, linked to each other and promoted by the mainstream media, may assume many of the functions and pacifying effects of mainstream corporate media. Excessive time in the blogosphere could detract from other activities, ranging from study to interpersonal interaction to political engagement. Small Internet communities may be deluded that they are the revolution and may miss out on participation in key social movements and struggles of the age. Yet any technology can be misused, abused, and overused; therefore, we have sought to chart reasons for the hope that emergent media ecologies also provide increased opportunities to situate an oppositional technopolitics within the context of existing political struggles and movements.

Conclusion: Situating Oppositional Technopolitics

The analyses in this chapter suggest how rapidly evolving media developments in techno-culture make possible a reconfiguring of politics and culture and a refocusing of participatory democratic politics for everyday life. In this conjuncture, the ideas of Guy Debord and the Situationist International are especially relevant with their stress on the construction of situations, the use of technology, and the creation of alternative and oppositional cultural forms to promote a revolution of everyday life aimed at the increase the realm of freedom, community, and empowerment.[11] To a certain extent, then, the emergent information and communication technologies are transformative in the direction of more participatory and democratic potentials. Yet it must be admitted that this progressive dimension coevolves with processes that also promote and disseminate the capitalist consumer society, individualism, and competition and that have involved emergent modes of fetishism, alienation, and domination yet to be clearly perceived and theorized (Best and Kellner, *The Postmodern Adventure*).

Emergent media ecologies are thus contested terrains produced by tools used by the Left, Right, and Center of both dominant cultures and subcultures in order to promote their own agendas and interests. The political battles of the future may well be fought in the streets, factories, parliaments, and other sites of past struggle, but politics is already mediated by broadcast, computer, and other information technologies and will increasingly be so in the future. Our belief is that this is at least in part a positive development that opens radical possibilities for a greater range of opinion, novel modes of virtual and actual political communities, and original forms of direct political action. Those interested in the politics and culture of the future should therefore be clear on the important role of the alternative public spheres and intervene accordingly, and critical cultural theorists and activists have the responsibility of educating students concerning the techno-cultural and subtechnocultural literacies that ultimately amount to the skills that will enable them to participate in the ongoing struggle inherent in the present age (Kellner, "Technological Transformation").

Technopolitics have become part and parcel of the mushrooming global movement for peace, justice, and democracy that has continued to grow through the present and shows no sign of ending. The emergent movements against capitalist globalization have thus placed squarely before us the issue of whether or not participatory democracy can be meaningfully realized. Whereas the mainstream media had failed to vigorously debate or even report on globalization until the eruption of a vigorous anticapitalist globalization movement and rarely, if ever, critically discussed the activities of the WTO, World Bank, and the International Monetary Fund (IMF), there is now a widely circulating critical discourse and controversy over these institutions. Whereas prior to the rise of the recent antiwar, prodemocracy movements average citizens were unlikely to question a presidential decision to go to war, now people do question and not only question but protest publicly. Such protest has not prevented war nor successfully turned back globalized development, but it has continued to evoke the potential for a participatory democracy that can be actualized when the public reclaims and reconstructs technology, information, and the spaces in which they live and work.

Online activist subcultures and political groups have thus materialized in the last few years as a vital oppositional space of politics and culture in which a wide diversity of individuals and groups have used emergent technologies to help produce creative social relations and forms of democratic political possibility. Many of these subcultures and groups may become appropriated into the mainstream, but no doubt novel oppositional cultures and different alternative voices and practices will continue to appear as we navigate the increasingly complex present toward the ever-receding future.

Emergent media ecologies such as the Internet provide the possibility of an alternative symbolic economy, forms of culture and politics, and instruments of political struggle. It is up to the oppositional groups that utilize these digital tools to develop the forms of technopolitics that can produce new relations of freedom and can liberate humanity and nature from the tyrannical and oppressive forces that currently constitute much of our global and local reality. The present challenge for critical theorists of emergent media and the cyberculture is to begin to conceive their political reality as a complex network of places embodying reconstructed models of citizenship and novel forms of political activism, even as the emergent media ecologies reproduce logics of domination and become co-opted by hegemonic forces. In this sense, media and political activists should especially look to how emergent digital tools and cultures are interacting as tentative forms of self-determination and control "from below" (Marcuse 180–82), recognizing that today's mediated citizen-activists represent historical oppositional forms of agency in the ongoing struggle for social justice and a more participatory democracy.

Notes

1.   Early discourses of cyberoptimism include Rheingold, *The Virtual Community: Homesteading on the Electronic Frontier*; Barlow, "A Declaration of the Independence of Cyberspace"; Gates, *The Road Ahead*; and Kelly, *Out of Control*. By "technopolitics" we mean politics that is mediated by the wide range of digital tools such as broadcasting technology, cameras, the hardware and software involved with computers, multimedia, and the Internet. Thus Internet politics and a myriad of other forms of media politics are contained under the more general concept of technopolitics, which describes the nature of the proliferation of technologies that are engaged in political struggle. In this paper, while we speak broadly about the innovative developments occurring between the Internet, other new media, and general populations, we will look specifically at how new World Wide Web forms are influencing and being influenced by technopolitics and culture. For more on technopolitics, see Kellner ("Intellectuals, the New Public Spheres, and Technopolitics"), Armitage, (*New Cultural Theory and Technopolitics*), and Best and Kellner (*The Postmodern Adventure*). For useful comments that helped us significantly in revising this article, we would like to thank the editors of this collection.

2.   See Dean, "Communicative Capitalism." While we agree with Dean's analysis of the potential fetishism of the Internet and technopolitics, we believe that at times she illicitly totalizes the concept of fetishism in her denunciation of Internet cultural politics. Dean's recent book on the subject, *Publicity's Secret: How Technoculture Capitalizes on Democracy*, does a more well-rounded job of valorizing the positive political potentials of the Internet and her idea concerning self-organizing networks of contestation and struggle is one that we are very

much in support of here. While it is beyond the scope of this paper to engage in more detail her polemic on the public sphere, we would argue for a more dialectical vision that seeks consensus and commonality amidst differences, but which emphasizes struggle and difference as a check to hegemonic consensuality, as well as to counter radically anti-democratic forces like the Bush administration.

3.   Extreme right-wing material is also found in other media, such as radio programs and stations, public access television programs, fax campaigns, video and even rock music productions. Republicans began using new technologies and intervening in alternative media seriously during the Reagan years (see Kellner, *Television and the Crisis of Democracy*).

4.   Ivan Illich's "learning webs" (*Deschooling Society*) and "tools for conviviality" (*Tools for Conviviality*) anticipate the cross-development of the Internet and other new media and how they may provide resources, interactivity, and communities that could help revolutionize education. For Illich, science and technology can either serve as instruments of domination or progressive ends. Hence, whereas big systems of computers promote modern bureaucracy and industry, personalized computers made accessible to the public might be constructed to provide tools that can be used to enhance learning. Thus, Illich was aware of how technologies like computers could either enhance or distort education depending on how they were fit into a well-balanced and multidimensional ecology of learning.

For a pre-Internet example of the subversion of an informational medium, in this case, public access television, see Douglas Kellner, "Public Access Television: Alternative Views" (*Television and the Crisis of Democracy*), also available at http://www.gseis.ucla.edu/courses/ed253a/MCkellner/ACCESS.html (accessed 26 Feb. 2006). Selected episodes are freely available for viewing as streaming videos at http://www.gseis.ucla.edu/faculty/kellner (accessed 26 Feb. 2006).

5.   "Blogs" are hypertextual Web logs that people use for new forms of journaling, self-publishing, and media/news critique, as we discuss in detail later. People can implement their blog via software engines such as Blogger (http://www.blogger.com), Moveable Type (www.moveabletype.org), Typepad (http://www.typepad.com), and Live Journal (http://www.livejournal.com). It was estimated that there were some 500,000 blogs in January 2003, while six months later the estimated number claimed to be between 2.4 and 2.9 million with ten million estimated by 2005; see http://www.blogcensus.net for current figures. For examples, see our two blogs: BlogLeft: Critical Interventions, http://www.gseis.ucla.edu/courses/ed253a/blogger.php, and Vegan Blog: The (Eco)Logical Weblog, http://www.getvegan.com/blog/blogger.php. "Wikis" are popular new forms of group databases and hypertextual archives, covered in more depth later in this paper.

6.   See Delio. Another Iranian blogger, Hossein Derakhshan, living in exile in Toronto develops software for Iranian and other bloggers and has a popular Web site of his own. Hoder, as he is called, worked with the blogging community to launch a worldwide blogging protest on July 9, 2003 to commemorate the crackdown by the Iranian state against student protests on that day in 1999 and to call for democratic change once again in the country. See his blog at http://hoder.com/weblog. On recent political blogging in Iran, see Thomas.

7.   See http://www.dailykos.com (accessed 26 Feb. 2006) for an example.

8.   For information, resources, and groups within the Media Reform Movement, see http://www.freepress.net.

9.   See http://www.iht.com/articles/126768.html (accessed 26 Feb. 2006) for an example of how the World Economic Forum, while held in increasingly secure and remote areas, has been penetrated by bloggers.

10.   For elaboration of this argument, see Kellner's *Media Spectacle and the Crisis of Democracy* and his "Digital Technology and Media Spectacle," at http://www.gseis.ucla.edu/faculty/kellner/essays/digitaltechnologymediaspectacle.pdf.

11.   On the importance of the ideas of Debord and the Situationist International to make sense of the present conjuncture, see chapter 3 of Best and Kellner, *The Postmodern Turn*, and on the new forms of the interactive consumer society, see Best and Kellner, *Postmodern Adventure*.

Works Cited

Armitage, John, ed. Special Issue on Machinic Modulations: New Cultural Theory and Technopolitics. *Angelaki: Journal of the Theoretical Humanities* 4.2 (September 1999).

Barlow, John Perry. "A Declaration of the Independence of Cyberspace." *Barlow Home(stead) Page.* 1996. July 2003 http://www.eff.org/~barlow/Declaration-Final.html.

Best, Steven, and Douglas Kellner. *The Postmodern Adventure*. New York and London: Guilford and
    Routledge, 2001.
———. *The Postmodern Turn*. New York and London: Guilford and Routledge, 1997.
Bowers, C. A. *Educating for Eco-Justice and Community*. Athens: University of Georgia Press, 2001.
Brecher, Jeremy, Tim Costello, and Brendan Smith. *Globalization From Below*. Boston: South End, 2000.
Burbach, Roger. *Globalization and Postmodern Politics: From Zapatistas to High-Tech Robber Barons*. London:
    Pluto, 2001.
Castells, Manuel. "Flows, Networks, and Identities: A Critical Theory of the Informational Society." *Critical
    Education in the New Information Age*. Ed. D. Macedo. Lanham, Md.: Rowman and Littlefield, 1999.
    37–64.
Couldry, Nick, and James Curran, eds. *Contesting Media Power: Alternative Media in a Networked World*.
    Boulder, Colo., and Lanham, Md.: Rowman and Littlefield, 2003.
Cvetkovich, Ann, and Douglas Kellner. *Articulating the Global and the Local: Globalization and Cultural
    Studies*. Boulder, Colo.: Westview, 1997.
Dean, Jodi. "Communicative Capitalism: Circulation and the Foreclosure of Politics." *Cultural Politics* 1.1
    (March 2005): 51–74.
———. *Publicity's Secret: How Technoculture Capitalizes on Democracy*. Ithaca: Cornell University Press, 2002.
Delio, Michelle. "Blogs opening Iranian Society?" *Wired News* 28 May 2003. http://www.wired.com/culture/
    lifestyle/news/2003/05/58976.
Dyer-Witheford, Nick. *Cyber-Marx: Cycles and Circuits of Struggle in High-Technology Capitalism*. Urbana:
    University of Illinois Press, 1999.
Feenberg, Andrew. *Alternative Modernity*. Los Angeles: University of California Press, 1995. 144–66.
———. *Questioning Technology*. London and New York: Routledge, 1999.
Gates, Bill. *The Road Ahead*. New York: Penguin, 1996.
Hafner, Katie. "For Some, the Blogging Never Stops." *New York Times* 27 May 2004.
Illich, Ivan. *Deschooling Society*. New York: Harper, 1971.
———. *Tools for Conviviality*. New York: Harper, 1973.
Jenkins, Henry, and David Thorburn. *Democracy and New Media*. Cambridge, Mass.: MIT Press, 2003.
Jordan, Javier, Manuel Torres, and Nicola Horsburgh. "The Intelligence Services' Struggle Against al-Qaeda
    Propaganda." *International Journal of Intelligence and Counterintelligence* 18.1 (Spring 2005): 31–49.
Kellner, Douglas. *From 9/11 to Terror War: Dangers of the Bush Legacy*. Lanham, Md.: Rowman and
    Littlefield, 2003.
———. "Globalization and the Postmodern Turn." *Globalization and Europe*. Ed. R. Axtmann. London:
    Cassells, 1998.
———. *Grand Theft 2000: Media Spectacle and a Stolen Election*. Lanham, Md.: Rowman and Littlefield,
    2001.
———. "Intellectuals, the New Public Spheres, and Technopolitics." *New Political Science* 41–42 (Fall 1997):
    169–88.
———. "The Media and the Crisis of Democracy in the Age of Bush-2." *Communication and
    Critical/Cultural Studies* 1.1 (2004): 29–58.
———. *Media Spectacle and the Crisis of Democracy*. Boulder, Colo.: Paradigm Publishers, 2005.
———. "Technological Transformation, Multiple Literacies, and the Re-visioning of Education." *E-Learning*
    1.1 (2004): 9–37.
———. *Television and the Crisis of Democracy*. Boulder, Colo.: Westview, 1990.
———. "Theorizing Globalization." *Sociological Theory* 20.3 (Nov. 2002): 285–305.
Kelly, Kevin. *Out of Control*. New York: Perseus, 1995.
———. *New Rules for the New Economy*. New York: Viking, 1998.
Leuf, Bo, and Ward Cunningham. *The Wiki Way: Collaboration and Sharing on the Internet*. Boston: Addison-
    Wesley, 2002.
Luke, Allan, and Carmen Luke. "A Situated Perspective on Cultural Globalization." *Globalization and
    Education*. Ed. N. Burbules and C. Torres. London and New York: Routledge, 2000.
Luke, Carmen. "Cyber-Schooling and Technological Change: Multiliteracies for New Times." *Multiliteracies:
    Literacy, Learning and the Design of Social Futures*. Eds. B. Cope and M. Kalantzis. Australia: Macmillan,
    2000. 69–105.

Marcuse, Herbert. "The Historical Fate of Bourgeois Democracy." *Towards a Critical Theory of Society: Collected Papers of Herbert Marcuse*. Ed. D. Kellner. Vol. II. London and New York: Routledge, 2001.

McCaughey, Martha, and Michael Ayers. *Cyberactivism: Online Activism in Theory and Practice*. London and New York: Routledge, 2003.

McLuhan, Marshall. *Understanding Media: The Extensions of Man*. Cambridge, MA and London: The MIT Press, 2001.

Meikle, Graham. *Future Active: Media Activism and the Internet*. London: Taylor and Francis, 2002.

Newitz, Annalee. "The Trouble with Anonymity on the Web." *AlterNet*. August 20, 2007. http://www.alternet.org/columnists/story/60298/.

Nie, Norman H., and Lutz Ebring. *Internet and Society: A Preliminary Report*. Stanford, Calif.: The Institute for the Quantitative Study of Society, 2000.

Orlowski, Andrew. "Google to Fix Blog Noise Problem." *The Register* 9 May 2003. http://www.theregister.co.uk/content/6/30621.html.

Poster, Mark. "Cyberdemocracy: The Internet and the Public Sphere." *Internet Culture*. Ed. D. Porter. London and New York: Routledge, 1997. 259–72.

Rheingold, Howard. *Smart Mobs: The Next Social Revolution*. Cambridge, Mass.: Perseus, 2002.

———. *The Virtual Community: Homesteading on the Electronic Frontier*. Reading, Mass.: Addison-Wesley, 1993.

Steger, Manfred. *Globalism: The New Market Ideology*. Lanham, Md.: Rowman and Littlefield, 2002.

Sunstein, Cass R. *Republic.com*. Princeton, N.J.: Princeton University Press, 2002.

Thomas, Luke. "Blogging Toward Freedom." *Salon* 28 Feb. 2004. http://archive.salon.com/opinion/feature/2004/02/28/iran/index_np.html.

Trend, David. *Welcome to Cyberschool: Education at the Crossroads in the Information Age*. Lanham, Md.: Rowman and Littlefield, 2001.

Turkle, Sherry. *Life on the Screen: Identity in the Age of the Internet*. New York: Touchstone, 1997.

Van Alstyne, Marshall, and Erik Brynjolfsson. "Electronic Communities: Global Village or Cyberbalkans?" *Proceedings of the International Conference on Information Systems*. 1996. http://aisel.isworld.org/password.asp?Vpath=ICIS/1996&PDFpath=paper06.pdf.

Winner, Langdon. "Citizen Virtues in a Technological Order." Eds. E. Katz, A. Light, and W. Thompson. *Controlling Technology*. Amherst, N.Y.: Prometheus, 2003. 383–402.

4

Remembering Dinosaurs: Toward an Archaeological Understanding of Digital Photo Manipulation

Karla Saari Kitalong

> She paints with layers of images, piling them up, stripping them back—partially, fully, adding again. What she ends up with is distinctive and beautiful, and in a manner of its process, archaeological.
>
> —Christine A. Finn, *Artifacts*

With this description of artist Julieanne Kost's Photoshop composition, Christine A. Finn makes an intriguing connection between the art of digital photo manipulation and the science of archaeology. Finn is not the first to develop an archaeological metaphor for a nonarchaeological kind of activity; Sigmund Freud made a similar comparison between the work of the archaeologist and that of the psychotherapist, as this chapter illustrates. And Michel Foucault's archaeological strategy for historical interpretation has emerged as a method of understanding culture, power, and knowledge. Both metaphors capitalize on layered representations that characterize archaeology's method for understanding space and time.

Because they encourage the juxtaposition of dissimilar objects, concepts, or structures, metaphors and analogies have great explanatory power. They afford unique understandings, facilitate teaching and learning, and infuse color and excitement into language. In this chapter, I amplify what was for Finn a brief passage about the "archaeological" photo manipulation artistry of one of the people she met while conducting an ethnographic study of contemporary Silicon Valley. Inasmuch as the cultural role of photo manipulation is itself complexly layered—indeed, archaeological—in its ethical, legal, and technological implications, I propose that critically reframing photo manipulation through an archaeological metaphor can help us make sense of its complexity as a tool and as a cultural force. I argue, in common with William J. Mitchell, that the emergence of digital imaging affords an opportunity to expose certain contradictions in "photography's construction of the visual world" as well as to "deconstruct the very ideas of photographic objectivity and closure" (Mitchell 7). I push the metaphor as far as it will go and close with an assessment of where it takes us.

Archaeology and the Art and Science of Photo Manipulation

Like the archaeologist who reconstructs a way of life by assembling and analyzing a collection of excavated artifacts, Julieanne Kost[2] creates her art by assembling "numerous and seemingly disparate visual cues" (Finn 40). Kost begins each new project with a blank screen. Like the archaeologist whose workspace is an unexcavated landscape, Kost doesn't know precisely what the blank screen will yield. Nonetheless, she is confident that "[s]omething will . . . appear" (Finn 41). But neither the trained digital artist nor the professional archaeologist simply starts working, willy-nilly, hoping that something relevant will emerge. An archaeologist begins with some kind of material evidence—the trace of a campfire, a stray artifact, the remains of a rubbish heap or a roof beam—and a theory about the lifeways of an ancient civilization. Kost's artistic inspiration is not precisely equivalent to archaeological evidence, but like the archaeologist's impetus to excavate, Kost's impetus to create often comes from a found object that evokes an emotion, calls up a memory, or suggests a connection. Ultimately, her art grows out of a desire or need to communicate a message or express a mood. "I am not looking for the perfect photo," Finn quotes, "but anything that causes an emotional reaction" (41). The digital artist, then, begins with artifacts, a blank screen, and an idea. Finn writes, "Julieann [Kost] takes me through an image—'Surgery Flat'—created using Adobe Photoshop™. It is a multiplicity of layers—tractor tread, moss, a woman's back—actually Julieanne's to represent her mother's immobilized vertebrae—more organic material, sand dollars" (41).

To achieve the desired visual effect, Kost appropriates artifacts from a variety of contexts—a forest, a human body, a beach—and assigns a digitized version of each artifact to its own Photoshop layer. Because the layers are transparent, and because they are made up of pixels instead of molecules, she can employ as many layers as she wants or needs, subject of course to the limitations of her computer. And because each layer can be adjusted individually, composing becomes, in effect, a three-dimensional act of manipulation. Kost builds her images by adding and removing layers and by positioning, repositioning, and resizing the objects delegated to those layers. In this way, she experiments with different groupings, different emphases, and different emotions.

Moreover, as she adds, removes, reconstitutes, blends, and flattens the layers that comprise her works of art, she also compiles a kind of archaeological site on her computer's hard drive, a site within which her work ultimately is contextualized. The technology, in short, records Kost's movements in an archive. "There is a history of her palette," writes Finn. "She can go back to older images—back in creative time—and recall how she achieved effects, and how the layers of color and texture and meaning come together to describe her abstract impressions" (41).

An archaeological view of photo manipulation practices operates, then, on at least two levels: The artist builds her images in layers, and she assembles an "archaeological site" on her computer's hard drive. Cognitive psychologist Ulrich Neisser writes, "Out of a few stored bone chips, we remember a dinosaur" (285).[3]

When Freud noticed a connection between his avocation—an interest in famed archaeological discoveries such as the Rosetta Stone and the ruins of Pompeii—and his vocation—psychotherapy—he began to develop an archaeological metaphor for the theory and practice of psychotherapy. He was especially intrigued by archaeology's "belief in stratification" and its "desire for reconstruction" (Hake 148). He noticed that, like archaeologists, psychotherapists begin their work with a virtually blank landscape—a patient's unexplained problem—which they characterize by studying the patient's symptoms. Armed with theories of human cognition and emotion, the therapist begins to "excavate" like an archaeologist—to probe into the patient's layered memories, thoughts, and dreams with the goal of unearthing a plausible explanation for the patient's symptoms. When it is successful, psychotherapy, too, "remembers a dinosaur"; that is, it mobilizes fragments of memory and restores what is missing in order to concretize and thereby explain troublesome neuroses (Hake 152). It is this connection between archaeological and psychotherapeutic excavation on which Freud's followers, including Neisser, constructed an archaeological theory of psychotherapy.

From Freud's Archaeology of Psychotherapy to an Archaeological View of Photo Manipulation

Steen F. Larsen articulates the features of an archaeological theory of psychotherapy in such a way as to help us explore photo manipulation as an archaeological practice. He first observes that "memories have at least 'kernels of truth,'" or "immutable facts that have been preserved from the past," which he sees as analogous to the artifacts buried in the strata of an archaeological site (188). Likewise, images manipulated with Photoshop contain kernels of the original image from which they are composed.[4]

Larsen also points out that excavated objects or facts must be contextualized with respect to debris from other times—other strata—that accumulates around and on top of them (188). Such debris obscures the artifacts before they are unearthed; moreover, removal of the debris and exposure of the artifacts may also disclose new associations. For example, because of erosion, earthquakes, and other environmental factors, a 500-year-old arrowhead may end up just inches away from a soda bottle cap. In psychotherapy, analogously, the Freudian slip (such as substituting "mother" for "other") is thought to reveal unconscious associations that are viewed as potentially revelatory. In psychotherapy and archaeology as well as in photo manipulation, then, when relationships that cross strata are noticed, analyzed, and manipulated, new meanings may emerge.

A Memorial Day essay published in the *Milwaukee Journal Sentinel* provides a fitting example of how this phenomenon crosses over into the world of photo manipulation. In the essay, retired journalist Sue Diaz tellingly recounts her process of making a composite photograph as an assignment for her Photoshop class. She started with two images of uniformed soldiers taken some fifty years apart. In the older photograph, her father stood with his arm around her mother, and in the more recent one, her son squinted into the camera. The resulting composite photo blended the two soldiers into one image, the World War II–era grandfather's arm now draped over his Gulf War–era grandson's shoulders. "And just like that, he holds his grandson in an embrace that reaches beyond the boundaries of time" (Diaz). As she worked with her composite photo, Diaz began to see a new relationship between her father and her son, one that could not exist outside the virtual reality of the Photoshop image, because Diaz's father had passed away when his grandson was only about five years old.

> The merged picture continues to take shape, and I realize that the connection between these two men goes beyond the toolbar of Photoshop. With every click of the mouse, my sense of this grows stronger. The uniform each wears joins them in a bond, a brotherhood I can't possibly comprehend. What do I know of war? (Diaz)

Larsen's next point, "Remembering is a process of excavating such remainders of the original experience" (188), suggests that if we have enough exposure to recast images like Diaz's composition of the two soldiers, they may augment and even replace original memories in our consciousness. Although she alone was responsible for the creative work that generated the composite photograph, as soon as her creation began to take shape, it took on a life of its own: Diaz ultimately felt excluded from the exclusive bond between soldiers that the image exemplified.

Larsen also points out that it is necessary for psychotherapists who employ an archaeological metaphor in their practice to "clear . . . away from the memory objects any irrelevant stuff that adheres to them and prevents them from re-presenting the past in its original apparition" (188). In Photoshop, the trim, lasso, and crop tools serve this function: Users of photo manipulation software determine the relevance of each element of a photo, decide which elements to remove, and control the arrangement of the remaining elements. The ability to accomplish this selection process depends in part upon technical prowess and practice, upon tool skill. Technically, an ex-spouse can be removed from a family photograph and the edges blended to erase any hint of her presence. But deleting the face from the image does not delete the person from memory. Residue of the relationship—the kernel of truth—remains, although we may choose not to see or acknowledge it.

Larsen points out, as well, a truism of life in a visual age: "The excavated memory objects will usually be only fragments of the original state of affairs" (188). Photographs, whether or not they have been digitally manipulated, are intrinsically mere fragments of a larger event. In photographs that have been digitally recombined or manipulated, further fragmentation occurs before the image is finally reconstituted once again. "From these fragments," Larsen adds, "the rememberer (or the analyst) must try to reconstruct a full picture of the past, using whatever additional intuitions, knowledge, and theory he has" (189). Neisser's phrase, "Out of a few stored bone chips, we remember a dinosaur," aptly captures this reconstructive version of archaeological practice, regardless of the domain of knowledge-making in which the metaphor is applied. In the absence of an unbroken clay pot, a first-hand experience, or an original image, the past tends to become even more fragmentary; the original exists only in individuals' memories.[5]

Technicalities: The Archaeology Metaphor and Photoshop's Layer Tools

Before the concept of layers was introduced to Photoshop in version 3.0, prompting Russell Rohde to call the software the "8th wonder of the world" (17), Photoshop artists relied upon "floating selections"—artifacts that had been selected, copied, and pasted into an image. Floating selections could be moved or resized as long as they remained selected.

Once deselected, however, the imported images merged with their new context. Under such conditions, writes Rohde, "Rescuing an original image requires use of 'undo' (when possible), 'revert' or having foresight to have created a 'duplicate' which remains unaltered. Only one floating selection can be done at a time" (17).

The layers tool affords much more control over a composite image than did the floating selection function. Once the concept of layers had been added to Photoshop, in fact, it became the software's defining process. Ultimately, a later innovation, adjustment layers, permitted the artist to add a "special layer of correction information" that "only affects those layers below it" in the image. These "special purpose layers store color and tone information, but no image information" (Ashbrook 19). Like other layers, adjustment layers can be reordered, deleted, or adjusted without making changes to the image as a whole. Only when the layers are flattened does the image "stay put."

Indeed, the introduction of various permutations of the layers tool opened the door for further development of an archaeological frame of mind. Thus, in Photoshop today, artifacts and their attributes—color, shininess, contrast—can be separated, each in its own layer, to maximize the artist's control over the meaning conveyed by the image. Similarly, archaeologists (and psychotherapists who understand their practice as archaeological) aggregate previously unknown artifacts to build explanations of people's daily lives.

The "careful and knowledgeable investigator" tries to minimize the disturbance he or she causes to buried material or conceptual objects (Larsen 188). In both archaeology and psychotherapy, though, once the landscape has been disturbed, the memories disrupted, there is no turning back. Both archaeology and psychotherapy destroy as they recover, preserve, or recuperate: An archaeologist must disturb a landscape if he or she expects to examine artifacts and thereby come to understand the previous inhabitants' lives. When archaeologists—deliberately or unwittingly—excavate sacred burial grounds, the destruction is doubled.[6] Similarly, psychotherapy, while usually thought of as a healing practice, may unearth repressed issues that dissolve relationships and even destroy lives. You can't undig a hole or unprobe an emotion.

In contrast, artist Kost is permitted—indeed, encouraged—to adjust, transform, and mutate the artifacts with which she composes, either individually or as a group, both before and after she incorporates them into her composition. Photoshop's crop, lasso, and cut tools are designed to destroy digitized images in the service of creating. However, the destruction wrought by photo manipulation is only figurative: "The component parts are played with, teased, softened, textured until the artist is satisfied" (Finn 41). But the original photographic artifacts remain intact after they have been digitized; in addition, as noted earlier in this chapter, consecutive versions of the digitized artifacts can be preserved as a kind of archaeological site on the artist's hard drive. Kost can capture as many versions of her composition as she wants; "[b]eing digital, the image can be stored and restored over time" (Finn 41). Assuming that the artist's computer is equipped with enough disk space, all versions of the image can be retained, as can the original visual artifacts that comprise it. Even the deleted "scraps" of a digitized image can be saved and later restored.

In short, the creative work of photo manipulation, while bearing certain resemblances to archaeology and to an archaeological understanding of Freudian psychotherapy, still retains the undo function, the archive of the previous version, the intact original, begging the question, So what? Of what use is this extended metaphor? To address this question, we must situate photo manipulation within a larger cultural context, twenty-first century technoculture, where we can begin to see that artistic photo manipulation, while instructive for teasing out the metaphor's details, is relatively inconsequential in the larger debate surrounding photo manipulation.

Historical Truth vs. Artistic Splendor: The Extent and Limits of an Archaeological Metaphor for Photo Manipulation

Larsen concludes his discussion of the archaeology metaphor for psychoanalysis by comparing professional with amateur archaeology: "The archaeology metaphor implies clear, practical guidelines for carrying out 'memory excavations' by analogy to the way archaeological excavations are supposedly done." He goes on to contrast the work of "amateur archaeologists" who "dig only for precious objects that catch the eye, tickle the imagination, or are wanted by fanatical collectors" with that of "professional, scientific archaeologists," who, he says, proceed quite differently because their aim is historical truth and not splendor. They search every inch of the field of excavation with equal care and attention; they record the features of any object that comes to light; they systematically consider the significance of each object, however unremarkable it may appear; and they carefully preserve the objects for later scrutiny (Larsen 189).

Up to this point, I have been discussing photo manipulation strictly as artistic work, as a creative endeavor, a comparison that works well in relating archaeological excavation, with its emphasis on layers, to image construction. My discussion has not yet touched on what many consider to be the dark side of photo manipulation. It is a "short step," Mitchell observes, "from innocuous enhancement or retouching to potentially misleading or even intentionally deceptive alteration of image content." The malleability of a digital photograph calls into question the distinction between "the objective, scientific discourses of photography and the subjective, artistic discourses of the synthesized image" (Mitchell 16). The archaeological metaphor for photo manipulation gives way, then, precisely at the point where "historical truth" gives way to "splendor."

A recent journalism ethics case provides an interesting illustration of this metaphorical slipperiness (Sreenivasan). In April 2003, Brian Walski, an award-winning *Los Angeles Times* photojournalist, was covering the early weeks of the war in Iraq. One of his photographs subsequently appeared on the front pages of several prominent American newspapers, including the *Los Angeles Times* and the *Hartford Courant* (Irby, par. 3). The photograph was viewed, of course, as documentary evidence of a moment in the war. Because of public expectations that journalists are objective, that journalists convey the truth, any embellishment of the truth is viewed as deceptive.[7] In fact, Walski had deliberately embellished by digitally fusing the left half of one photograph to the right half of another to achieve a more dramatic composition (Irby).

Walski's photographic fusion was relatively insignificant as a technological act; in fact, it was barely noticeable. Irby explains that the composite nature of Walski's photograph was discerned only after it had already been published, when an employee at the *Courant* noticed an anomaly—an apparent "duplication in the picture"—and decided to magnify the image to take a pixel-level look. The employee brought it to the attention of the copy desk, which then immediately alerted *Courant* Assistant Managing Editor Thom McGuire. "After about a 600 percent magnification in Photoshop, I called Colin [Crawford, *Los Angeles Times* director of photography] to ask for an investigation," McGuire says (Irby, par. 8).

According to a number of critics, the doctored image renders suspect all of Walski's previous accomplishments. Wren asks, "Knowing what we know now about his photo in the *LA Times*, aren't we almost required to question the authenticity of these clever juxtapositions in his other photos?" An analogy can be drawn to archaeology: The Society for American Archaeology (SAA) decries what it calls "the commercialization of archaeological objects," defined as "their use as commodities to be exploited for personal enjoyment or profit" (Society, par. 5). Such commercialization "results in the destruction of archaeological sites and of contextual information that is essential to understanding the archaeological record" (Society). Furnishing one's own curio cabinet or fattening one's wallet by selling found objects from archaeological sites, then, ignores a basic tenet of archaeological practice, that the "archaeological record" is "irreplaceable" and must be protected. SAA therefore mandates the professional archaeologist to conduct "responsible archaeological research" (Society). Similar codes govern the conduct of psychotherapists and photojournalists (British; Lester).

Because he crossed the line—rather than "preserv[ing]" an event "for later scrutiny," Walski tried to achieve "splendor" (Larsen 189)—he lost not only his job but also his professional credibility. His actions also contributed to a growing public distrust of photojournalism. According to Mitchell, "The inventory of comfortably trustworthy photographs that has formed our understanding of the world for so long seems destined to be overwhelmed by a flood of digital images of much less certain status" (19). Walski acknowledged his transgression of photojournalistic norms in a brief apology to his colleagues, but insisted that it was a one-time lapse. "I have always maintained the highest ethical standards," he wrote, "and cannot explain my complete breakdown in judgment at this time" (qtd. in Van Riper). Frank Van Riper adds, "Walski's action will call into question the work of any number of other reporters and photographers."

Photoshop, Archaeology, and the Nature of the Text

Despite the stated commitments to objectivity and ethical practice inscribed in their professional codes of ethics, photojournalism, psychotherapy, and archaeology can all be understood as reconstitutive activities: Psychotherapists use cognitive theories to help patients reconstruct repressed memories in much the same way as archaeologists apply theories of material culture to reconstruct lost civilizations. Photo manipulation tools likewise allow—even encourage—users to construct new visualizations. Photoshop (the tool) and photo manipulation (the practice) make it easy to juxtapose artifacts from different temporal and spatial contexts, to excise unwanted artifacts, to reconstitute memories; that is, photo manipulation changes the nature of photographic practice. Brian Walski is not the first professional photographer to manipulate a photograph in order to shape the meanings available to audience members. *Time* magazine darkened O. J. Simpson's mug shot, *Newsweek* replaced Bobbie McCaughey's teeth, and *National Geographic* "repositioned" the pyramids, to name just three of the more controversial and visible cases

(Sreenivasan; Lester; Mitchell). These journalists join Julieanne Kost, Sue Diaz, and Brian Walski in the work of designing our memories—of "remembering dinosaurs" on our behalf.

Photo manipulation is more than a technical skill; questions such as, How can I get a compelling image out of this series of mediocre photographs? must be accompanied by more complex questions, not the least of which are, Should I try to create a more compelling image? and What are the cultural consequences of doing so? Neisser writes, "The same fragments of bone that lead one paleontologist to make an accurate model of an unspectacular creature might lead another, perhaps more anxious or more dramatic, to 'reconstruct' a nightmarish monster" (97–98). Indeed, despite the SAA's eighth principle, which mandates that archaeologists "ensure that they have adequate training" (Society, par. 10), implying continued attention to theoretical developments, two archaeologists working with the same artifacts may draw on different theoretical perspectives. Similarly, because of divergent theoretical emphases, two psychotherapists studying the same patient's symptoms may produce alternative explanations—and prescribe completely different treatments—for the same neurosis. But psychotherapists' and archaeologists' interpretations are constrained by the boundaries of acceptable practice required for licensure. Acceptable practice for artists, on the other hand, mandates that they produce wildly divergent interpretations; thus, it is safe to say that, faced with the same set of artifacts from which to choose and the same Photoshop tools, no two artists would create exactly the same composition. Moreover, each artist's work—his or her selection and manipulation of visual elements within Photoshop—can be expected to lead both the artist and the viewer to a new understanding.

Thus, it seems to me that in addition to questioning the practice of photo manipulation, we must also, as Foucault would have us do, question the document; that is, we must ask "not only what these documents [mean], but also whether they [tell] the truth, and by what right they could claim to be doing so" (6). In other words, we must consider the cultural positioning not only of artifacts produced in the practice of photo manipulation but also the cultural positioning of the practice itself. This entails questioning cultural "divisions or groupings with which we have become so familiar," thereby suspending the unities that we take for granted (Foucault 22). Among these groupings, these unities, is the assumption that we can define and recognize photojournalistic truth.

Photojournalists occupy a cultural position somewhere in between art and science. Their profession mandates that they report objectively—as Van Riper puts it, "*news photographs are the equivalent of direct quotations* and therefore are sacrosanct" (emphasis in original). At the same time, in order to make an impact and ultimately to sell newspapers, their photographs must be visually compelling, even striking. The cultural presence of photo manipulation software intensifies this paradox. Again, Walski's case provides an apt illustration. Individually, Walski's images are discrete texts, but as a group, the series of images he made of the engagement between a British soldier and Iraqi civilians also constitute a text—a recording of a series of moments on a given day and time, in a given place. Whether we take as our text the individual photographs, the series of photographs, or the conflated representation of the event constructed after the fact from available artifacts, texts are necessarily fragments, just as Larsen describes memories as "only fragments of the original state of affairs" (188).

With an archaeological perspective on photo manipulation such as I have outlined in this chapter, however, we can see beyond the surface of the discrete image to see the layers, the strata, the discontinuities that comprise it. For Walski, such seeing included reconciling the competing forces of technological capability, aesthetic value, and journalistic integrity. From an assemblage of layered, manipulated, altered, and often discontinuous artifacts, then, Photoshop users manufacture continuities that deliver specific, albeit individually interpretable, messages.

Photo manipulation itself, however, persists as a discontinuity, a cultural "floating selection" not yet fused to its context. The messages formed by photo manipulation express truths—or propagate lies—that might otherwise be inexpressible. They deliver meanings that cannot otherwise be articulated. A soldier points his gun at startled civilians, a grandfather joins his grandson in a photograph of two soldiers, the pyramids are made to appear more beautiful and exotic, and a mother's back pain is rendered visible so that others can vicariously experience it. The collage is complete.

Notes

1.   References to Photoshop in this chapter serve as "shorthand" for the entire range of photo manipulation software. Photoshop is a widely known and used example of this class of products.

2.   For more about Kost and samples of her work, see http://www.adobeevangelists.com.

3.   Such kernels of truth formed the basis for a humorous Adobe Systems ad campaign that ran several years ago to introduce a new version of Photoshop. In the campaign, families were depicted using Photoshop to redeem snapshots they didn't like: In one "episode," a teenager's garish hairdo was restyled for the family Christmas card; in another, an ex-spouse was seamlessly excised from a reunion photo.

4.   This is what John Berger, in *Ways of Seeing*, calls "mystification."

5.   Modern archaeological practice insists on reinterring excavated bodies, albeit after research has been done.

6.   I am indebted to Ed Scott for alerting me to this story as it was emerging and for writing about it in our Visual Texts and Technologies doctoral seminar in April 2003.

7.   Photojournalists hold themselves to this standard; see the National Press Photographers' Association Code of Ethics (Lester, Appendix A), especially tenets 2 and 3.

8.   The two original photographs and the resulting composite can be viewed online at http://www.sree.net/teaching/photoethics.html. The front pages of the *Los Angeles Times* and the *Hartford Courant* and a flash animation comparing original and fused photographs can be viewed online at Poynter Online, the online version of the Poynter Institute, a journalism school, http://www.poynter.org/content/content_view.asp?id=28082.

Works Cited

Ashbrook, Stanley. "Photoshop's Adjustment Layers." *PSA Journal* 65.8 (August 1998): 19–21.

Berger, John. *Ways of Seeing*. New York: Penguin, 1977.

British Association for Counseling and Psychotherapy. "Ethical Framework for Good Practice in Counselling and Psychotherapy." 1 April 2007. 20 August 2007.
    http://www.bacp.co.uk/ethical_framework/ethical_framework_web.pdf.

Diaz, Sue. "Memorial Day: Both Warriors, But Generations Apart." *Milwaukee Journal Sentinel* 25 May 2003, 01J.

"Digital Manipulation Code of Ethics." *National Press Photographer's Association*. 1995. 27 Apr. 2004 http://www.nppa.org/professional_development/business_practices/digitalethics.html.

Finn, Christine A. *Artifacts: An Archaeologist's Year in Silicon Valley*. Cambridge, Mass.: MIT Press, 2001.

Foucault, Michel. *The Archaeology of Knowledge and the Discourse on Language*. New York: Pantheon, 1972.

Hake, Sabine. "Saxa Loquuntur: Freud's Archaeology of the Text." *Boundary 2* 20.1 (1993): 146–73.

Irby, Kenny F. "L.A. Times Photographer Fired Over Altered Image." *Poynter Online* 12 Oct. 2003 http://www.poynter.org/content/content_view.asp?id=28082.

Larsen, Steen F. "Remembering and the Archaeology Metaphor." *Metaphor and Symbolic Activity* 2.3 (1987): 197–99.

Lester, Paul Martin. *Photojournalism: An Ethical Approach*. Hillsdale, N.J.: Lawrence Erlbaum, 1991. http://commfaculty.fullerton.edu/lester/writings/pjethics.html.

Mitchell, William J. *The Reconfigured Eye: Visual Truth in the Post-Photographic Era*. Cambridge, Mass.: MIT Press, 1992.

Neisser, Ulrich. *Cognitive Psychology*. New York: Appleton-Century-Crofts, 1967.

Rohde, Russell A. "Photoshop—8th Wonder of the World." *PSA Journal* 62.6 (June 1996): 17.

Scott, Ed. "Deception and Omission: Unfortunate Tools for Photojournalists in a Digital Age." Unpublished paper, University of Central Florida, 2003.

Sreenivasan, Sreenath. "Famous Examples of Digitally Manipulated Photos." *Photo Ethics* 27 Apr. 2004 http://www.sree.net/teaching/photoethics.html.

Society for American Archaeology. "Principles of Archaeological Ethics." 2005. *Archaeology for the Public*. 20 August 2007. http://www.saa.org/public/resources/ethics.html.

Van Riper, Frank. "Frank Van Riper on Photography: Manipulating Truth, Losing Credibility." *Washington Post* 9 Apr. 2003. 2 June 2005 http://www.washingtonpost.com/wpsrv/photo/essays/vanRiper/030409.htm.

Wren, Paul. "Don't Fool with Mother Nature." *Lunch with George!* 9 June 2003. 20 August 2007. http://www.lunchwithgeorge.com/lunch_030403.html.

5

Cut, Copy, and Paste

Lance Strate

For almost all of the past three million years, "high tech" took the form of tool kits consisting of sharpened pieces of stone (see, for example, Leaky and Lewin). These tool kits included three types of stone knives: one for skinning and otherwise preparing meat, another for working with vegetables and fruits, and a third used to create other tools out of wood and bone. The technique of chipping stones in order to create a sharp cutting edge might have been developed by the australopithecines, but it was regularly used by *Homo habilis* and subsequent members of our genus such as *Homo ergaster, Homo erectus, Homo heidelbergensis, Homo neanderthalensis,* and ultimately early *Homo sapiens.* The tool kits were ubiquitous in our prehistory to an extent that is truly extraordinary, as they are found across a vast range of climates and habitats, across isolated populations separated by great distances, and most remarkably, across different species of hominids. The stone tools themselves were subject to surprisingly little innovation or evolution over this long period of time, and the creative explosion that laid the foundations for modern technological societies did not occur until approximately thirty thousand years ago (Pfeiffer).

The distance from the stone chips of the hominids to the silicon chips of home computers seems unimaginable. And yet it is all but instantaneous in comparison to the millions of years during which the prehistoric tool kit served as the indispensable key to the survival and success of genus *Homo.* Moreover, these two moments of stone and silicon are connected by a "missing link"—the technologies of writing. The stone knife used to sharpen, carve, and fashion other tools out of wood and bone also could make marks on their surfaces, both accidental and purposeful, both aesthetic and symbolic. Notches, engravings, inscriptions, and other markings constitute some of the main forerunners of writing and other notational systems (Gelb; Schmandt-Besserat). Writing, in turn, supplies the basic concepts of code and of logical operations that undergird the computer revolution (Hobart and Schiffman; see also Strate, "Containers, Computers, and the Media Ecology of the City"). Given the continuity between the prehistoric and the cybernetic and their close proximity from the point of view of deep time, it is possible to recognize the residue of the Stone Age in contemporary culture, as Joshua Meyrowitz does when he describes us as "hunters and gatherers in an information age" (315). More than a metaphor, what Meyrowitz is describing is the retrieval of previously obsolesced behavior patterns, a process described by Marshall and Eric McLuhan as one of the laws of media.

Meyrowitz probably did not have the hunt-and-peck mode of typing in mind when he wrote about contemporary hunting and gathering, but the fact of the matter is that most computer technology relies heavily on keyboards and on information in the form of electronic text. And just as our ancestors used stone tools to hunt, gather, and process their food, we use digital tools to hunt and gather alphanumeric data and to process words, numbers, images, and sounds. Of course, we have a great many tools in our digital tool kits, but there is, I would argue, a contemporary equivalent to the three stone knives of prehistory: cut, copy, and paste. I would go so far as to say that these three commands are archetypes of word processing and computer programming and that they are possibly ideal, certainly exemplary forms of the human activity of tool use. As such, the basic function of cut, copy, and paste is the modification and manipulation of human environments. These digital tools influence and alter our information environment, much as the prehistoric tool kit was used to act on the natural environment. Of the three, "cut" is clearly a direct descendant of the stone knife. The icon used to represent cut is a pair of scissors, which is, after all, a pair of knives joined together to facilitate cutting. But cutting only requires a sharp object, one that could be found rather than made, so the act of cutting would predate the development of a cutting tool such as the stone knife. As the most ancient of activities, it is fitting that cut appears first in the edit menu of current operating systems.

Copy, at first glance, appears to be entirely modern, a product of "the age of mechanical reproduction," to invoke Walter Benjamin's (217) classic study. But the act of copying predates its mechanization and consequent standardization via printing technology (Chaytor; Eisenstein; Steinberg). Scribal copying of manuscripts was characterized by corruption and multiformity, but the result was still multiple copies of a text. Folk artists are incapable of producing artifacts that are exactly identical, but what they do produce is nonetheless formulaic and imitative and thus a series of copies. There is no verbatim memorization in oral cultures, which means that every oral performance may be unique; however, singers of tales do sing the same songs over and over again, so each performance can be seen as a copy of an earlier one. The ubiquity of the stone knives themselves over millions of years is a testament to the primacy of the act of copying. No doubt cutting came first, through the use of found objects. But copying implies the deliberate manufacture of cutting tools, not just tool use but toolmaking. It represents a great leap forward in technological development as well as biological evolution, a milestone on the road to humanity.

As for "paste," it is a tool whose traces are wiped out over time, in contrast to the stone knives that speak to us across millions of years, and it is therefore difficult to discuss pasting with any certainty. However, it seems reasonable to conclude that our evolutionary ancestors noticed the adhesive properties of certain substances and employed them in limited ways prior to the creative explosion. Paste, therefore, is appropriately listed last of the three in the edit menu, and its development may well have been based on the processing of animal and vegetable products that stone knives made possible.

Whatever their origins, we can certainly conclude that cutting, copying, and pasting originate as prehistoric technical activities, as methods of manipulating the environment.

Moreover, these three tools cover the most basic types of technological operations: Cutting implies taking a whole and breaking it into smaller parts. Pasting implies joining together separate parts to form a larger whole. And copying implies the repetition of either or both of the other two actions in order to achieve similar results. We can also see here the beginnings of arithmetic operations: Cutting represents subtraction in that you take part away from the whole; insofar as the parts that are left do not disappear but remain as parts of the whole, cutting also represents division. Pasting represents addition in that you join parts to create a whole. And copying, of course, represents multiplication. Interestingly, this technological perspective diverges from the standard view of arithmetic, in which addition is seen as the most basic operation, followed by subtraction, then multiplication, and finally division. Instead, we can understand that it is subtraction and division that are the most natural of operations, the closest to the human lifeworld—to get to addition and multiplication we need to develop a higher level of technology. Similarly, the logic of computer programming also differs from standard arithmetic logic in significant ways (i.e., the separation of data and operations).

Of course, we can represent and subsequently manipulate our environment with the aid of words as well as numbers, and the technology of language contains the equivalent of the cut, copy, and paste tools. For example, in describing an experience, we cut or leave out information that we consider irrelevant or undesirable, we paste or elaborate on and embellish our accounts, and we copy or repeat a story to different audiences and on different occasions. Language allows us to edit reality, but language itself is difficult to edit as long as it remains locked in the form of speech. It is not until the introduction of the written word that it becomes possible to fix language in space. The translation of speech into visual form allows us to step back and distance ourselves from our own words and those of others, to view and review language, and ultimately to criticize and revise our words. Writing begats rewriting, in the process of which words can be added, moved, removed, or copied. Cut, copy, and paste remain implicit and subsumed under the general activity of editing during the early history of writing, but the three basic functions allow for the editing of linguistic reality, the secondary editing of a reality already edited by the spoken word. In sum, the basic function of cut, copy, and paste is the modification and manipulation of symbolic environments.

From writing to printing to digital media, the technologizing of the word, to use Walter Ong's phrase, has enhanced our editing functions. The digital tool kit has allowed us to improve upon our linguistic creations, to take our written texts and make them better (Perkinson, "Getting Better," "How Things," "No Safety"). As language gives us the basic idea of perfection (Burke), subsequent technologies of the word reduce our tolerance for imperfection, altering our expectations and therefore our perceptions and conceptions of perfection (Gumpert; Strate, "The Conception of Perfection"). Thus, cut, copy, and paste give rise to new standards of perfection, and we are only beginning to recognize those standards as they emerge out of the digital media environment. Certainly, we become less forgiving of typographical errors in the era of electronic text and more likely to demand rewrites of students and subordinates. Perfectionism can be counterproductive, of course, when it gives rise to endless revision on the part of the writer. Rearranging text becomes so easy with cut, copy, and paste that it may overwhelm the act of composition itself. Michael Heim argues that "the immediacy of composing on computers can make even a small degree of frustration an excuse for writers to turn from composing to picayune editing" (154). Heim does have a point here, although the characterization of electronic editing as picayune seems overly harsh. We might instead think of it as *hyper-editing*, which implies that text is infinitely rearrangeable and therefore anticipates *hyper-text*. In other words, the hypertext format is the ultimate expression of cut, copy, and paste.

Hypertext represents one type of freedom, the freedom of multidimensionality experienced by the reader (Strate, "Cybertime," "Hypermedia"). Another type of freedom engendered by cut, copy, and paste is the freedom experienced by the writer working in an infinitely rearrangeable medium. In this regard, the digital tool kit has altered the process of composition. Previously, composition had been largely a matter of mouths and minds. In an oral culture, composition is inseparable from performance (Lord); simply put, you make it up as you go along, albeit working from memory and a tradition of formulaic expressions. When writing was introduced, it was understood to be a method to record utterances and correct mistakes in the transcription; composition was essentially a process of dictation, either to a scribe or to oneself. As writing became internalized, composition became a direct expression of thought as much as it was a recording of speech (Nystrom). The advent of silent writing paralleled that of silent reading, and compared to older methods of composition grounded in orality, this more direct process probably suggested the possibility of an automatic writing that was associated with an entirely different type of *medium*. But the internalization of writing typically results in some form of prewriting, that is, composing sentences within one's head, which precedes the act of putting pen to paper or fingers to typewriter keys. The digital tool kit, on the other hand, allows for a different, more direct form of composition, as Heim notes: "The thought process previously called prewriting can be carried out on screen because the physical impressions of handwriting and typewriting are removed from the immediate work environment. When fluidity is emphatic and graphic, the psyche feels that nothing is written in stone. The preliminary phases of composition are less likely to daunt the person beginning to set down some initial thoughts. What is entered as the first sentence on a word processor need not be considered first in any permanent or logical way. Automated editing functions . . . generate quite a different sense of the risk involved in committing oneself to writing" (153).

Just as the introduction of writing itself freed the mind from the need for memorization (Ong), word processing frees the mind from the need for prewriting. Thus, cut, copy, and paste contribute to a purer form of written composition. Electronic editing makes it possible to employ more spontaneous and immediate forms of writing, dispensing, at least for the moment, with any effort to tailor the text to a particular audience (Heim), which is a figment of the writer's imagination anyway (Ong). And while Heim sees this as characteristic of only word processing, it would also explain some of the unusual dynamics of e-mail communication and listserv interaction.

The significance of cut, copy, and paste is not limited to written composition, however. Just as one of the stone knives in the prehistoric tool kit was used to create other tools from wood, bone, and other materials, the technology of writing is used to develop innumerable other technologies, electronic editing is used to edit computer programs in codes as diverse as machine language and HTML. The digital tool kit can also alter databases. There really is no formal distinction between traditional written documents and binary code, although the semantic and pragmatic differences are vast. In using commands like cut, copy, and paste to edit computer programming, we are using digital tools to create and change other digital tools. In this sense, one of the basic functions of cut, copy, and paste is the modification and manipulation of technological environments.

Still, it is not sufficient to consider text and code, word and number, as the raw materials for the digital tool kit. Another basic function of cut, copy, and paste is the modification and manipulation of visual environments. The introduction of writing represents the translation of an acoustic medium, speech, into a visual symbol system. The printing press placed even greater stress of visual arrangement of text and information and on the aesthetics of layout and typographic space (Eisenstein; Ong). Printing, of course, provided a highly standardized form of copying, whereas the use of moveable type is related to cut and paste procedures. But it was the more recent innovations of offset printing and xerography in the twentieth century, both representing further advances in copying, that led to the common practice of "cut and paste" layout, utilizing actual scissors and glue. This is not to dismiss the artistic practice of collage, which precedes and anticipates cut and paste procedures, but collage is associated with a unique artwork, while "cut and paste" is used to create an original for mechanical reproduction. Moreover, documents produced by cut and paste methods are meant to be transparent and immediate, indistinguishable from more traditional forms of printing; in contrast, collage intentionally calls attention to its construction and is therefore hypermediated (see Bolter and Grusin). Through cut and paste procedures, typewritten text and handmade illustrations, along with art "clipped" or cut out of other print media, could be pasted onto master documents for easy reproduction, democratizing the production of print media. The experience of cutting and pasting together flyers, newsletters, and the like and then taking them to copy centers to be reproduced was fairly widespread until recently, even for elementary school children. Of course, cut and paste procedures could also be used for purely written documents as a form of editing, with the expectation that the seams would not show after photocopying. But "cut and paste" was truly revolutionary because it gave amateurs control over layout, the visual organization of text, images, and so on, allowing for the creation of documents resembling professionally produced print advertisements and periodicals.

The translation of cut and paste operations to the computer's electronic environment involved both automation and etherealization (as the object being manipulated was no longer made from paper but instead binary code). And this allowed cut, copy, and paste to become purely writing tools, as noted earlier. But their basis is in visual organization and reorganization, and they are essential tools for all forms of desktop publishing and digital design. For that matter, they are vital to any form of computer graphics production. We therefore can conclude that cut, copy, and paste have a visual bias. Their visual bias is the source of their editing power, but it also relates to their strong connection to the graphical user interface (GUI). This triad of editing tools represents one of the most ubiquitous features of the GUI, which is best known in the form of the Macintosh and Windows operating systems (Barnes). The GUI not only makes the three tools easily available but also makes them easily memorizable as part of an invariable system of menus. As Steven Johnson observes in *Interface Culture*, "Every Mac user knows how to cut and paste because he or she knows *where* the copy and paste commands are—in the upper-left-hand corner of the screen, under the 'Edit' menu item. The knowledge becomes second nature to most users because it has a strong spatial component to it, like the arrangement of letters on a QWERTY keyboard. As with the original typewriter design, the consistency of the layout is as important as the layout itself" (77).

Only the file menu commands appear more frequently (e.g., quit, open, close, print, and save), and these commands do not actually process data or modify and manipulate information but simply deal with programs and data sets as wholes. The edit menu follows immediately after the file menu (moving from left to right across the top of the screen) and therefore is arguably the first true set of tools that the GUI makes available to the computer user. Other commands may be found under this menu (e.g., undo, find, and replace), but cut, copy, and paste are *always* found there, always together as the primal set of digital tools. It should be noted that this editing triad predates the GUI. The digital tool kit has its australopithecine era as command line instructions used in early editing programs. But it is the GUI, like genus *Homo*, which standardizes the tool kit and unlocks its power. The mouse plays a vital role here, as it is used to highlight and select the text and objects that the digital tools will act upon.

In turn, cut, copy, and paste unleash the power of the GUI. A great deal of attention has rightfully been given to the development of the GUI as a major breakthrough in computer technology (Barnes). But the focus has mostly been on the use of icons, the desktop metaphor, and the mouse, and few have noted the role that the editing triad has played in establishing the practical utility of the GUI. Cut, copy, and paste are the key to the desktop metaphor and the defining activity of cycling through windows (Turkle). For example, I might use a word processing program to compose an e-mail message, editing it to eliminate the errors and ambiguities typical of more casual e-mail communications; once I am satisfied with the composition, I select the text, copy it from the word processing program, and paste it into the e-mail program. I might also copy text from e-mail messages and paste them into word processing documents to edit and save the information. Or I might copy URLs and other information from the Web and paste it into e-mail messages or word processing documents. I could copy and paste data, text, and objects among database, spreadsheet, desktop publishing, and multimedia applications. Or I might copy an image from a clip art file and paste it into a document. I can also call up my calculator application, perform an arithmetic operation, and copy and paste the answer into a text file.

The common ground in all these maneuvers is the ability to move data, text, or objects from one program to another without resorting to a complicated set of export and import commands. Through the aid of a hidden fourth tool, the "clipboard," not only can I cycle through windows but I can also cut, copy, and paste across the windows. The simplicity and speed of these operations is what makes the GUI so powerful. Without cut, copy, and paste, the windows on the desktop are isolated islands; the editing triad bridges these islands. Moreover, the range of programs that utilize cut, copy, and paste is quite remarkable. Apart from text, data, code, and graphics, the editing triad is a fixture in programs devoted to audio and video files, for example. The distinctions between these different forms and media are irrelevant, since they all are nothing more than sets of data—bits rather than atoms, as Nicholas Negroponte likes to say in *Being Digital*. The digital tool kit modifies and manipulates binary code, and although we typically do not deal with a bunch of zeroes and ones, this code is the universal language that the GUI is built on.

Therefore, like the GUI itself, cut, copy, and paste reflect the biases of binary code. Lev Manovich notes this underlying logic in *The Language of New Media*:

> As an example of how the interface imposes its own logic on media, consider "cut and paste" operations, standard in all software running under the modern GUI. This operation renders insignificant the traditional distinction between spatial and temporal media, since the user can cut and paste parts of images, regions of space, and parts of a temporal composition in exactly the same way. It is also "blind" to traditional distinctions in scale: the user can cut and paste a single pixel, an image, or a whole digital movie in the same way. And last, this operation renders insignificant the traditional distinctions between media: "cut and paste" can be applied to texts, still and moving images, sounds, and 3-D objects in the same way. (65)

In addition to being an expression of digital coding, Manovich also argues that cut, copy, and paste encourage new practices and establish new aesthetics related to collage, montage, and postmodern pastiche:

> The practice of putting together a media object from already existing commercially distributed media elements existed with old media, but new media technology further standardized it and made it much easier to perform. What before involved scissors and glue now involves simply clicking on "cut" and "paste." And, by encoding the operations of selection and combination into the very interfaces of authoring and editing software, new media "legitimizes" them. Pulling elements from databases and libraries becomes the default; creating them from scratch becomes the exception. The Web acts as a perfect materialization of this logic. It is one gigantic library of graphics, photographs, video, audio, design layouts, software code, and texts; and every element is free because it can be saved to the user's computer with a single mouse click.
>
> It is not accidental that the development of the GUI, which legitimized a "cut and paste" logic, as well as media manipulation software such as Photoshop, which popularized a plug-in architecture, took place during the 1980s—the same decade when contemporary culture became "postmodern." (130–31)

It follows that cut, copy, and paste are consistent with contemporary emphases on consumption, nostalgia, self-reference, hypermediacy, and the ironic mode. Rather than foster the creation of original media elements, the digital tool kit facilitates the purchase (or appropriation) of commercially distributed ones, thereby shifting the focus from production to consumption. Since already available elements would have to be older than newly made ones, there is also a shift from present to the past. Cut, copy, and paste make it easy to recycle (or rerun) older material, thereby evoking a sense of nostalgia, a longing for the past, but not necessarily a coherent image of the past in the form of a history. The juxtaposition of readymade objects from different times and places results in an emphasis on contrasting styles rather than the representation of reality. This has a self-referential quality to it, as the mixture of forms makes us aware of form itself and specifically of the cut and paste process. In that sense, the elements that are recycled lose their meaning and we focus on the medium and not any message it might seem to carry. The mixing of different media elements therefore destroys any immediacy or transparency in the communication process in favor of hypermediacy, the awareness of mediation (and remediation) as process (Bolter and Grusin). This absence of depth and meaning leaves room only for an ironic mode of expression, a sense that we ought not to take anything seriously since what we are witnessing is nothing more than the free play of existing elements.

Mark Twain once observed that when you have a hammer in your hand, everything around you starts to look like a nail. Along the same lines, we might imagine that for the early hominids holding stone knives in their hands, everything looks like something to cut, pierce, or carve. And for ourselves, armed as we are with the digital tool kit, do we not begin to see everything in our electronic environment as an object that can be cut, pasted, and especially copied? Clearly, cut, copy, and paste has contributed to the rise of plagiarism and the disregard for intellectual property rights characteristic of contemporary culture. For students, it has become all too easy to copy entire papers and articles from the Web and paste them into their documents (although search engines make it equally easy to catch the plagiarists). But more broadly, electronic text and the digital tool kit are challenging long-accepted notions of copyrights and intellectual property (Kleinman). In this regard the portable document format (PDF) serves what Paul Levinson (author of *Mind at Large* and *The Soft Edge*) refers to as a *remedial* function. Ostensibly a means of preserving format and layout over a myriad of different operating systems, browsers, and preference settings, the contents of PDF files cannot be easily copied and therefore are relatively immune to "cut and paste" operations and unauthorized editing, which is why they are sometimes preferred over text files. No doubt, there will be an increasing market for formats resistant to cutting, copying, and pasting as a means of balancing out the bias of the GUI.

Cut, copy, and paste tend to fade into the background, disappear into the edit menu, and become invisible and environmental, but that is exactly why they are so powerful. Our tools, which are so often inconspicuous and ubiquitous, form the deep structure of human cultures. From stone knives to spoken words to written composition to digital editing, we use our tools to modify and manipulate our environment, and our new environment, in turn, shapes and influences our own development. The implications of cut, copy, and paste extend well beyond the computer screen, mouse, and keyboard. These tools also have significance for biotechnology, suggesting recombinant DNA and cloning. They have significance for image politics, with its rearrangeable and interchangeable issues and candidates. They have significance for the production of consumer goods in an information economy, with its emphasis on branding and "on demand" production. They have significance for art, literature, and music in the postmodern era. They have significance for human identity and consciousness in the twenty-first century, with its increasing fragmentation, recombination, and fluidity. Cut, copy, and paste have significance for our future, which is why we need to understand our tools and the ways they are shaping us.

Works Cited

Barnes, Susan. B. "The Development of Graphical User Interfaces and Their Influence on the Future of Human-Computer Interaction." *Explorations in Media Ecology* 1.2 (2002): 81–95.

Benjamin, Walter. *Illuminations.* Trans. Harry Zohn. New York: Harcourt, 1968.

Bolter, Jay David, and Richard Grusin. *Remediation: Understanding New Media.* Cambridge, Mass.: MIT Press, 1999.

Burke, Kenneth. *Language as Symbolic Action: Essays on Life, Literature, and Method.* Berkeley: University of California Press, 1966.

Chaytor, Henry John. *From Script to Print: An Introduction to Medieval Vernacular Literature.* Cambridge: W. Heffer, 1950.

Eisenstein, Elizabeth L. *The Printing Press as an Agent of Change: Communications and Cultural Transformations in Early Modern Europe*. 2 vols. New York: Cambridge University Press, 1979.

Gelb, Ignace J. *A Study of Writing*. Rev. ed. Chicago: University of Chicago Press, 1963.

Gumpert, Gary. *Talking Tombstones and Other Tales of the Media Age*. New York: Oxford University Press, 1987.

Heim, Michael. *Electric Language: A Philosophical Study of Word Processing*. New Haven, Conn.: Yale University Press, 1987.

Hobart, Michael E., and Zachary S. Schiffman. *Information Ages: Literacy, Numeracy, and the Computer Revolution*. Baltimore: Johns Hopkins University Press, 1998.

Johnson, Steven. *Interface Culture: How New Technology Transforms the Way We Create and Communicate*. New York: HarperEdge, 1997.

Kleinman, Neil. "Striking a Balance: Access and Control, Copyright, Property, and Technology." *Communication and Cyberspace: Social Interaction in an Electronic Environment*. 2nd ed. Ed. L. Strate, R. L. Jacobson, and S. Gibson. Cresskill, N.J.: Hampton, 2003. 69–98.

Leaky, Richard E., and Roger Lewin. *People of the Lake: Mankind and its Beginnings*. New York: Avon, 1978.

Levinson, Paul. *Mind at Large: Knowing in the Technological Age*. Greenwich, Conn.: JAI, 1988.

———. *The Soft Edge: A Natural History and Future of the Information Revolution*. London: Routledge, 1997.

Lord, Albert B. *The Singer of Tales*. Cambridge, Mass.: Harvard University Press, 1960.

Manovich, Lev. *The Language of New Media*. Cambridge, Mass.: MIT Press, 2001.

McLuhan, Marshall, and Eric McLuhan. *Laws of Media: The New Science*. Toronto: University of Toronto Press, 1988.

Meyrowitz, Joshua. *No Sense of Place: The Impact of Electronic Media on Social Behavior*. New York: Oxford University Press, 1985.

Negroponte, Nicholas. *Being Digital*. New York: Alfred A. Knopf, 1985.

Nystrom, Christine. "Literacy as Deviance." *Etc.: A Review of General Semantics* 44.2 (1987): 111–15.

Ong, Walter J. *Orality and Literacy: The Technologizing of the Word*. London: Methuen, 1982.

Perkinson, Henry. *Getting Better: Television and Moral Progress*. New Brunswick, N.J.: Transaction, 1991.

———. *How Things Got Better: Speech, Writing, Printing, and Cultural Change*. Westport, Conn.: Bergin and Garvey, 1995.

———. *No Safety in Numbers: How the Computer Quantified Everything and Made People Risk-Aversive*. Cresskill, N.J.: Hampton, 1996.

Pfeiffer, John E. *The Creative Explosion: An Inquiry into the Origins of Art and Religion*. New York: Harper, 1982.

Schmandt-Besserat, Denise. *Before Writing: From Counting to Cuneiform*. Austin: University of Texas Press, 1992.

Steinberg, S. H. *Five Hundred Years of Printing*. Rev. ed. Ed. John Trevitt. New Castle, Del.: Oak Knoll, 1996.

Strate, Lance. "The Conception of Perfection: An Essay on Gary Gumpert and the Ambiguity of Perception." *Speech Communication Annual* 15 (2001): 96–113.

———. "Containers, Computers, and the Media Ecology of the City." *Media Ecology* 1 (1996): 1–13. http://raven.ubalt.edu/features/media_ecology/articles/96/strate1/strate_1.html.

———. "Cybertime." *Communication and Cyberspace: Social Interaction in an Electronic Environment*. 2nd ed. Ed. L. Strate, R. L. Jacobson, and S. Gibson. Cresskill, N.J.: Hampton, 2003. 361–87.

———. "Hypermedia, Space, and Dimensionality." *Readerly/Writerly Texts: Essays on Literature, Literary/Textual Criticism, and Pedagogy* 3.2 (1996): 167–84.

Turkle, Sherry. *Life on the Screen: Identity in the Age of the Internet*. New York: Simon and Schuster, 1995.

6

Dreamweaver and the Procession of Simulations: What You See Is Not Why You Get What You Get

Sean D. Williams

Metaphors are not culturally neutral; those that achieve circulation do so because they are consistent with the values of the culture.

—Tarleton Gillespie, "The Stories Digital Tools Tell"

In all aspects of life, not just in politics or in love, we define our reality in terms of metaphors and then proceed to act on the basis of the metaphors.

—George Lakoff and Mark Johnson,
Metaphors We Live By

Disneyland exists in order to hide that it is the "real" country, all of "real" America that *is* Disneyland.

—Jean Baudrillard, *Simulacra and Simulations*

What-you-see-is-what-you-get (WYSIWYG) software enacts a metaphor of simulation deeply rooted in American culture: we assume as Americans that "seeing is believing," that if we can only lay our eyes on an object or situation, our sight will penetrate to the true essence of the thing. Paradoxically, the metaphor of WYSIWYG software also reproduces a trait that informs American culture: we actively suspend our disbelief of what we see in order to make our lives easier. We know that vision frequently deceives us, yet because it's easier or more entertaining for us to simply ignore what we know, we consciously perpetuate—and perpetrate—acts designed to keep us in the dark. We choose not to see because it is easy and fun. Tarleton Gillespie, speaking about software's politics and its relationship to metaphor, suggests that metaphors don't evolve in a vacuum; they come from somewhere and that somewhere is the values of the culture itself (115).

As a consequence, our tools develop according to the metaphors that a culture finds acceptable: "seeing is believing" and "suspension of disbelief." George Lakoff and Mark Johnson, in their foundational study, *Metaphors We Live By*, examine the opposite side of the coin Gillespie tossed. Specifically, once a metaphor has become part of a culture, that culture's very reality is structured by the metaphor, and the way we act derives from the opposing sides of the metaphor (158). We go out of our way to "see" reality because seeing is believing, but we are perfectly content to ignore what we see—to suspend our disbelief—if it challenges our views or intellect. Jean Baudrillard, always the vigilant observer of America, comments on a similar phenomenon in *Simulacra and Simulation*, proposing that America defines the simulacrum, that we suspend our disbelief of our real values and priorities and instead choose to live according to a logic of simulation. This logic, expressed in Baudrillard's reference to Disneyland, hypothesizes that what we see is always a simulation. Disneyland simulates and represents some values of American culture, but at the same time, Disneyland's perspective on those values is so persuasive and appealing to us that we begin to replace "American values" with "Disneyland values" in a twist that makes Disneyland's representation of America stand in for the real America (Baudrillard 12). We accept what Disneyland offers as real because we see those values in action—seeing is believing, which requires us to suspend our disbelief of the obvious artifice. We accept the mediation Disneyland offers as a metaphor for America, and through the metaphorical simulation, America actually becomes what Disneyland offers. In an endless procession of the simulacrum, we see simulations of simulations without any need of reference to the real because the real has become what the simulacrum represents: Disney represents America, and then America reflects the simulation back to Disney, which Disney then re-represents as America, which then becomes America again. Disneyland both represents America and is the real country of America.

WYSIWYG software, like Disneyland, enacts America at the same time it reflects values already circulating. In fact, anytime a person sits down to use a computer, they are, in fact, enacting the logic employed by WYSIWYG software—the logic of simulation—since every interaction with a computer transacts through a mediated, and therefore metaphorical, structuration of language. Put differently, computers don't understand people and most people don't understand computers—except when those interactions pass through a filter that transforms natural language into a computer's language or transforms computer language into natural language. Every computer task works at this level of simulation, and we assume that what we see—or rather, read—from the computer screen is what really happens within the computer: WYSIWYG. Paradoxically, at the same time, we know that the computer can't really talk, think, or express emotions in happy-faced icons, yet we suspend our disbelief of this reality and prefer to anthropomorphize our machines.

The significance is great. Computer use has proliferated in the United States (and most industrialized countries) to the point that digital literacy's importance parallels that of reading, writing, and math. According to Internet World Stats, 69 percent of North Americans consider themselves frequent users of the Internet, while, globally, Internet use has increased from approximately twenty-three thousand users in 1995 to almost 1.2 billion in summer of 2007 (http://www.internetworldstats.com/stats.htm).[1] Given this ubiquity of computing in America—indeed the world—we must begin to reflect on our tools and how they impact culture (Gillespie 103). Tools are created by somebody to do something in particular ways; therefore, we need to examine the tool to show how it reflects— and creates—a view of the world just as we should think critically about the America that Disneyland proffers. To paraphrase Steven Woolgar, tools themselves—Adobe/ Macromedia's Dreamweaver in this case—seek "to define, enable, and constrain" (69). Tools are artifacts, Langdon Winner reminds us, that have politics (122): Dreamweaver, like all WYSIWYG software, is a product of the material world that has been designed, modified, and created for human action. Like Disneyland's reflected values that become the basis for action in our culture, Dreamweaver reflects back to us a perspective on working with digital texts that, in turn, becomes the basis for our future interactions with those texts. Therefore, we need to theorize the impacts of technology upon our culture; we need to discern how the tools themselves both represent and come to structure our culture. Or, in Lev Manovich's words, we need to articulate the relationships between the "computer layer" and the "cultural layer" (46) since both layers impact and shape each other in a process of simulation.

A Theory of WYSIWYG Software: Simulacra and Simulation

In *Simulacra and Simulation*, Baudrillard argues that the simulacrum is a cycle of continuous reference to the simulacra, or of reference to a representation, which comes to stand for the real through the procession of the simulacra. There are endless simulations of simulations without any need of reference to the real because the real has become what the simulacrum represents. The simulacrum is "the generation by models of a real without origin or reality: a hyperreal" (Baudrillard 1). The target of this chapter, Dreamweaver, embodies the simulacrum. Dreamweaver allows users to design Web sites and Web pages without any reference to coding in hypertext markup language (HTML) or Javascript. In this case, the graphical interface of Dreamweaver allows users to model in rich visual form the underlying code, to see on the surface what the code underneath means. In other words, the visual interface is a model of the HTML code; it's a simulation. In Baudrillard's terms, the visual interface actually becomes a simulacrum because its referent, the HTML code, is itself a simulation of the binary code that a machine understands. The visual interface is a simulation of a simulation—a hyperreal object—because it models something that is itself without origin.

The simulacrum represented by Dreamweaver relies on our two metaphors, "seeing is believing" and "suspension of disbelief," in order to enact this simulacrum. Specifically, the users of the interface believe that how they design a page using Dreamweaver is how the page will actually appear, and as Baudrillard would argue, this is an example of confusing the real with the model (17). But as Baudrillard would also argue, we have no choice in the age of simulation but to let the model stand for the real and, in turn, become the real itself. Therefore, in this case, we have multiple layers of simulation: the public layer, or the Web page itself, models the WYSIWYG interface of Dreamweaver, which models the HTML and other coding of the Web page, which itself models the binary machine language, which itself models a flow of electrons—the "actual" reality. We see, finally, not the real thing—the flow of electrons—but rather the fifth level of simulation, the public Web page that users will accept as the real instance of communication. As Baudrillard might argue, we choose to ignore our knowledge that the WYSIWYG interface stands in for the HTML code, which itself stands in for the machine language and the flow of electrons, because we want to have the pleasure of experiencing our creations as they will appear. We want the satisfaction of completion, which Baudrillard suggests fulfills the cycle of simulation, where the physical body and technology merge and "technology is an extension of the body" (111). The physical reality becomes confused with the simulation in technology, and in order to experience pleasure, we accept the artifice as real. In the simulation, where we have displaced the physical flow of electrons with various levels of linguistic mediation, we have lost a dimension of the original materiality of our communicative act (Kress and van Leeuwen 90), but we simply choose to suspend our disbelief—to believe in the simulation—because it's convenient, easy, or pleasant. This cycle of repeated simulations, or references to representations, embodies what Baudrillard calls the "simulacrum."

The importance of Baudrillard's propositions about living the simulacrum and how Dreamweaver enacts the simulacrum become clear when we consider technology as an extension of our culture. Specifically, as Manovich argues, "Strategies of working with computer data become our general cognitive strategies. At the same time, the design of software and the human-computer interface reflects a larger social logic, ideology, and imaginary of the contemporary society" (118). If we accept Manovich's perspective that technology can stand for society itself and that WYSIWYG software enacts the simulacrum, then our society itself, as Baudrillard argues, has no basis in reality. We choose to sublimate our knowledge that what we see is not real. Our culture chooses to live ever more on the surface; we accept our vision as reality, all the while ignoring the gnawing sense that there is more to it than we see. Our reliance on WYSIWYG software symbolizes our culture's choice to believe in a world that doesn't exist, and paradoxically, our belief in the world that doesn't exist causes that world to become our reality. This instance of "teleaction," as Manovich calls it (165), allows users to manipulate and construct reality through representation; the mediated stands in for the real and becomes the real itself: the perfect instance of the simulacrum.

Metaphors We Live By in Dreamweaver

If Dreamweaver, and WYSIWYG software generally, signifies our society's enactment of the simulacrum, how specifically does it do so? Metaphors. Metaphors, by their nature, simulate reality by asking us to accept something in place of another, to understand one experience in terms of another thing (Lakoff and Johnson 3). At its most fundamental level, Dreamweaver, as WYSIWYG software, enacts the primary metaphors of "seeing is believing" and "suspension of disbelief" as argued previously. On a more practical and superficial level, though, these metaphors are reified through interface metaphors of the tool itself. Three of these interface metaphors, examined in this chapter, participate in the simulacrum by adding yet another layer of representation to provide users with a frame to operate the tool. The interface metaphors make sense of the Web page authoring experience by providing a coherent structure that correlates with our knowledge and experience of the physical world, yet doesn't replicate that experience (Lakoff and Johnson 63). We are able to operate in the interface world exactly because we understand that the interface simulates first the HTML and other code, then the machine's binary language, then the flow of electrons.

So what metaphors does Dreamweaver employ to guide our actions? In the following sections I address three metaphors that reify the simulacrum constructed in part by WYSIWYG software and comment on their implications, because metaphors make sense of the world by highlighting some things and obscuring others (Lakoff and Johnson 164). The metaphors appear from most obvious to least obvious:

- Desktop windows
- Desktop publishing
- Montage

Desktop Windows

In their article "Politics of the Interface: Power and Its Exercise in Electronic Contact Zones," Cynthia and Richard Selfe argue that the primary computer interface, the desktop, reproduces objects familiar to a white-collar, corporate audience, "replicating a world that they know and feel comfortable within" (486). This primary metaphor of the desktop extends itself into Dreamweaver, where the interface clearly reproduces the most significant features of the desktop model. Windows arguably represent the most identifiable feature of the desktop model. As we see in Figure 1, the interface is divided into at least three windows. Each window has a different function, and to the uninitiated user of Dreamweaver, all these options can be very daunting, especially on a large monitor, where seven or more windows can be open at the same time. The windows metaphor forms the foundation of WYSIWYG software, because through these windows we see what we create immediately manifested before us. Perhaps without windows and the idea that you look

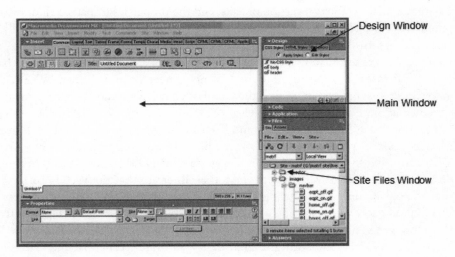

Figure 1. The windows of Dreamweaver

through a window onto something—in this case, your creations—WYSIWYG software would not be so popular.

Lakoff and Johnson invite readers to examine what metaphors obscure as well as what they elucidate, and in the case of windows, the metaphor suggests a pane between users and their creations, a physical distance between seer and seen. According to Manovich, this physical disruption of space and matter accompanies modernization and in doing so "privileges interchangeable and mobile signs over original objects and relations" (173). In other words, the window metaphor participates in the simulacrum by asking users to view the mediated reality—the thing seen through a window—as the object itself. Just as the power of all metaphors lies in their relationship or approximation to lived experience, the power of the simulation offered by windows resides in its simulation of our act of seeing reality through glass. We look at a window to watch the leaves blow on a maple tree and we see the real thing, the leaves blowing. By comparison, we look through a window in Dreamweaver and we see only representations—for example, windows, tabs, and file folders—that the metaphor asks us to accept as real.

Desktop Publishing

It probably seems a bit odd to suggest that "desktop publishing" (DTP) is a reality that Dreamweaver metaphorically recreates since DTP only has a twenty-five-year history. However, DTP grew in popularity along with the desktop computer itself such that nearly every computer user now has at least basic DTP capability available through preloaded software such as Microsoft Word or WordPerfect. While more powerful programs such as QuarkXPress, InDesign, and PageMaker dominate the professional DTP business, the "consumer" grade software reproduces many of the functions available in the professional packages. More importantly, Dreamweaver reproduces many of the functions present in DTP software at both the consumer and professional grades.

Compare, for example, the two screen shots in Figure 2, the first of PageMaker 6 and the second of Dreamweaver MX. PageMaker, distributed by Adobe, and Dreamweaver, originally distributed by Macromedia and now by Adobe, share many visual elements. Palettes, a main window, insertion objects, and the properties inspector all draw on a computer user's knowledge of DTP software for creating a Web page. Additionally, in more current programs like QuarkXPress, the newer versions of the Adobe products, and even in consumer grade software like Microsoft Word, you can actually create Web pages with your DTP software showing how the two fields have come together. WYSIWYG Web authoring tools like Dreamweaver, then, embody Jay Bolter and Richard Grusin's concept of remediation, taking the old model of DTP and transforming it for the new context of Web authoring (45). Dreamweaver and Web authoring tools in general try to replicate an older method of production for a new medium; Dreamweaver helps users make sense of Web authoring by drawing on a familiar gestalts of DTP (Lakoff and Johnson 86). The designers of Dreamweaver argue through their interface that the Web is nothing new and that the old models of producing text pages apply here as well. Web designers who use WYSIWYG software have no choice, of course, but to accept the metaphor offered.

However, in displacing direct access to HTML code with the ease and immediate gratification of WYSIWYG software, Web designers must accept that their decision carries cultural and social implications. Specifically, remediation, which is a metaphorical act of imposing an older communication model's paradigm and values on a newer medium

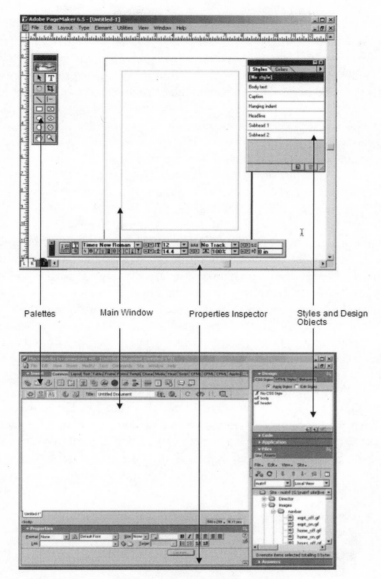

Figure 2.   Adobe PageMaker's interface vs. Macromedia Dreamweaver's interface

(imposing the print paradigm on the Web, in this case), allows users to build a coherent model of the software by "highlighting some things and hiding others" (Lakoff and Johnson 164). The DTP metaphor, then, participates in the simulacrum because it asks users to view their Web authoring experience in terms of another WYSIWYG experience based upon DTP software. Dreamweaver implements a WYSIWYG experience on top of another WYSIWYG experience and, in doing so, places the act of Web authoring squarely into the realm of simulation where Web design software simulates desktop publishing software, which is itself a simulation of typewriting technologies that simulate handwriting and ultimately mediate speech. Again, we see Dreamweaver enacting and relying on a procession of simulations to construct the "reality" of the Web authoring experience. Returning to Baudrillard, this procession of simulations defines the simulacrum.

Montage

According to Manovich in *The Language of New Media*, "collage" and "montage" historically dominate thinking about digital compositing, with montage—in which differences
between distinct objects are erased as they combine into a new product—now dominating
(124). In fact, in another instance of the simulacrum, the logic of montage reflects our current culture's burgeoning fascination with assemblage: taking ready-made objects and
splicing them together to create something useful for a context not imagined by the individual objects' creators (witness, for example, NASA's current practice of using "off-the-
shelf" technologies for space exploration machines). Rather than create things from
scratch, then, our current cultural context encourages assemblage; it asks us to rebuild and
recombine existing materials into novel configurations for novel purposes, all the while
blurring the lines between individual objects' original intentions or conceptions
(Manovich 131). This logic of recombination demonstrates Baudrillard's simulacrum or
cycle of endless reference because our new inventions become reflections of prior inventions. The reality of our new combinations only exists at the surface level of the novel configuration because the new combinations displace the original purposes of the individual
parts into a new whole that exists only in the novel configuration. There is not a "real"
behind the new assemblage—some solitary genius's blueprint for a fantastic machine that
the new configuration embodies. Instead, just like metaphors, the montage's origin resides
in a new configuration of raw materials already circulating in our culture; this demonstrates Baudrillard's "hyperreal," in which the surface structure becomes the reality and
subsequent basis for action and understanding.

WYSIWYG software, and Dreamweaver specifically, reifies this concept of "hyperreality" through the metaphor of montage. In most software like Dreamweaver, "every practical act involves choosing from some menu, catalog or database" demonstrating this
postmodern logic of assemblage (Manovich 124). Creating from scratch is the exception;
combining ready-made filters and objects is the norm. The obvious example shown in
Figure 3 is, of course, the menus themselves. Each menu contains predetermined ways of

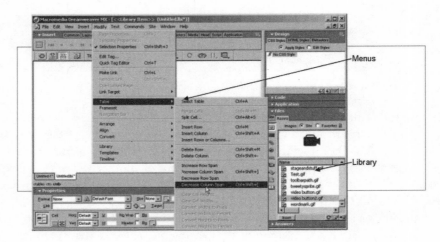

Figure 3.   The menus and library that demonstrate "assemblage"

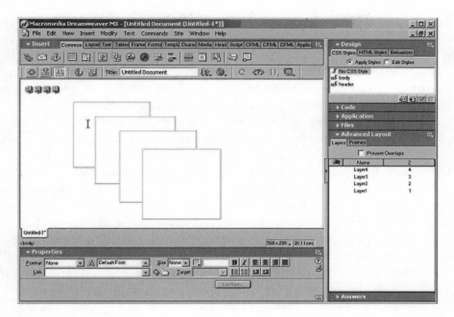

Figure 4.   Empty layers stacked

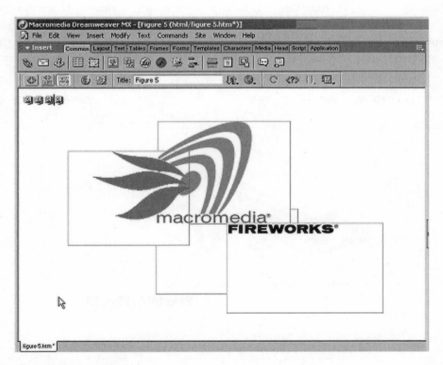

Figure 5.   Layers with content stacked

operating on the HTML code through the WYSIWYG interface. Also in Figure 3 we see that Dreamweaver (and actually all of Macromedia's digital communication development products, like Flash and Director) has a "library" (or, in the new version of Dreamweaver, an "assets" panel) where you can store ready-made "symbols" for inclusion on your Web site. These symbols are objects created by the user or others and placed into this database to be selected and then assembled at a later date into unique configurations.

Finally, through its "layers" option, shown in Figure 4, Dreamweaver encourages users to view content items as discrete elements to be configured and reconfigured. Designers can move, copy, and paste layers upon layers upon layers to achieve the desired visual effect. In Figure 4, for example, four empty layers have been placed on top of each other to abstractly demonstrate how layering works. Figure 5, by comparison, shows four layers stacked to create a single image, in this case a version of the logo for Macromedia's Fireworks program. These three elements, then—the menus, the library, and the layers— all embody the metaphor of montage by allowing users to assemble unique objects from discrete pieces to create a new product. What we see here is Baudrillard's "hyperreality," because there is no "original" production of artwork, only a simulation created by linking together previously discrete pieces into a new surface structure.

The object created by assembling the parts demonstrates a key aspect of the "montage" metaphor: compositing. According to Manovich, compositing comes natural to digital text because the base nature of all elements is the same thing—it's all numbers—and because digital text is modular (152). As a consequence of every object having the same nature and being viewed as a module in a system, it becomes very easy to stitch together pieces in such a way that their uniqueness becomes invisible. For example, when viewed in a Web browser, the layered image shown in Figure 5 becomes the seamless image of a Web page, shown in Figure 6. The image in Figure 6, then, demonstrates the hyperreal logic of compositing because it transforms the assemblage of Figure 5 into the flattened,

Figure 6.   Composited image

unified whole shown in Figure 6. The purpose of layers, in fact, is to create a composited, hyperreal montage, to splice together separate objects into a unified presentation that fools a user into seeing a single object rather than several discrete ones.

The logic of the montage returns us to the original paradox of "seeing is believing" and "suspension of disbelief" because the montage enacts the logic of the simulacrum, in which we choose to suspend our disbelief of difference and we choose to accept what we see rather than do the necessary work to distinguish among layers of simulations or separations among objects. In the montage metaphor, like the simulacrum itself, we're willing to accept the artifice of the construction where the pieces flow together into a seamless whole, even if that whole reveals internal contradictions or challenges our assumptions about what can coexist in a visual frame. In a fitting simulation of Baudrillard's simulacrum concept itself, the montage metaphor encourages us to *act* as assemblers working on the surface text but to *believe* that we create the original text by simultaneously enacting the primary metaphors of "seeing is believing" (after all, we appear to be creating something new and therefore believe we are) at the same time we suspend our disbelief of ourselves as assemblers (after all, we know we're actively compiling ready-made elements). Montage, finally, like the windows metaphor and the DTP metaphor, is a model that Dreamweaver exploits to generate "the real" experience of Web authoring, in which the reality is actually a hyperreality that relies on the linkages among models to generate "a real without origin or reality" (Baudrillard 1).

Conclusions

Examining the way that Dreamweaver privileges a hyperreal way of constructing digital text through the metaphors built into the software's interface seems fundamental not only for determining what the tool can help us accomplish (or prevent us from accomplishing) but also for understanding our culture itself. Johanna Drucker, a print artist speaking generally of technology's effect on writing, argues in *Figuring the Word* that writing is an equivalent of identity, so how we write—or in this case, how we author Web pages— reflects how we view ourselves and our place in the culture (229). Because crafting communication by its very nature lies at the intersection of personal and social expression, it can never be removed from the context in which it's embedded: "The cultural significance of a document . . . is hardly determined by its lexical content, but rather by the surrounding context in which the written text serves a critical part by its role as a document generating and being generated by the event" (Drucker 233). The software used to generate a document, or a Web page in our case, participates in the generation of the event; the software is part of the context that gives the communicative moment its meaning.

The metaphors used by Dreamweaver—windows, DTP, and montage—likewise contribute to the context of the communicative moment. We cannot separate context from the metaphors that build the context, and most importantly, we have to remember that "Metaphors are not culturally neutral; those that achieve circulation do so because they are consistent with the values of the culture" (Gillespie 115). The metaphors that we live by approximate our real, lived experiences, and consequently, we choose to live a life proffered by the tool's metaphors. The metaphors, therefore, come to stand for the real; the simulation is real. If we accept the reality of the simulation, the more important task becomes determining the implications of the simulation rather than trying to discover the real because, according to Baudrillard, that reality doesn't exist except in the simulacrum, or in the circulation and concatenations of simulacra—of continuous reference to representations, of continuous simulations of simulations. This chapter has suggested that Dreamweaver, by employing two paradoxical yet primary metaphors—"seeing is believing" and "suspension of disbelief"—and three subordinate metaphors—windows, DTP, and montage—participates in the simulacrum by reflecting back to us a perspective on digital text circulating in American culture that, in turn, becomes the basis for our future interactions with those texts. The reflection, or the story the digital tool tells, becomes a basis for structuring our reality, making a theorized understanding of the tool itself paramount for conceptualizing the nature of our social and cultural values. Dreamweaver and all WYSIWYG software, like Disneyland, enacts America at the same time it reflects values already circulating. In fact, anytime a person sits down to use a computer, they are, in fact, enacting the logic of the simulacrum because every interaction with a computer transacts through a procession of mediations and is therefore an interactive reality structured through simulations.

Note

1.   Determining the number of Internet users is not an exact science by any means. Some suggest that counting the number of servers on the Internet is the best method for determining usage, others suggest that traffic measurements are the best method, and still others suggest that the time users spend online is an indication of usage. Robert Hobbes's internet timeline is a good source for multiple indicators of usage, including the number of servers, hosts, networks, and domain names (http://www.zakon.org/robert/internet/timeline/).

Works Cited

Baudrillard, Jean. *Simulacra and Simulations*. Trans. Sheila Faria Glaser. Ann Arbor: University of Michigan Press, 1994.

Bolter, Jay David, and Richard Grusin. *Remediation: Understanding New Media*. Cambridge, Mass.: MIT Press, 2002.

Drucker, Johanna. *Figuring the Word: Essays on Books, Writing, and Visual Poetics*. New York: Granary Books, 1998.

Gillespie, Tarlton. "The Stories Digital Tools Tell." *New Media: Theories and Practices of Digitextuality*. Eds. Anna Everett and John T. Caldwell. New York: Routledge, 2003. 107–26.

Hobbes, Robert. "2000s." *Hobbes Internet Timeline*. 27 Aug. 2007 http://www.zakon.org/robert/internet/timeline/#2000s.

Internet World Stats. "Internet Usage Statistics—The Big Picture." 27 Aug 2007 http://www.internetworld stats.com/stats.htm.

Kress, Gunther, and Theo van Leeuwen. *Multimodal Discourse: The Modes and Media of Contemporary Communication*. London: Arnold Publishers, 2001.

Lakoff, George, and Mark Johnson. *Metaphors We Live By*. Chicago: University of Chicago Press, 1980.

Manovich, Lev. *The Language of New Media*. Cambridge, Mass.: MIT Press, 2001.

Selfe, Cynthia, and Richard Selfe. "Politics of the Interface: Power and Its Exercise in Electronic Contact Zones." *College Composition and Communication* 45.4 (1994): 480–504.

Winner, Langdon. "Do Artifacts Have Politics?" *Daedalus* 109.1 (1980): 122.

Woolgar, Steve. "Configuring the User: The Case of Usability Trials." *A Sociology of Monsters: Essays on Power, Technology, and Domination*. Ed. John Law. New York: Routledge 1991.

7

Revisiting the Matter and Manner of Linking in New Media

Collin Gifford Brooke

> To link is necessary; how to link is contingent.
>
> —Jean-François Lyotard, *The Differend*

Late in his career, Kenneth Burke, whose dramatistic pentad remains a staple of contemporary rhetorical and communications theories, remarked that one of his regrets was that he had not made his pentad a hexad. Burke explained that he would have added the term "attitude" (or "manner") to the five that most rhetoric students know by heart (i.e., agent, agency, act, scene, and purpose). Attitude, Burke reasoned, would lend additional nuance to the concept of agency, the term in the pentad that answers the question of how. In the context of a rhetorical theory (such as Burke's) that places a heavy emphasis on the psychological and nonrational elements of rhetoric, such nuance is warranted. When we ask how something has been accomplished, the answer is frequently one of attitude rather than functional agency; performing a task with gusto or with trepidation affects the character of the act as surely as the tool or tools one uses.

Much of the scholarship on technology might be also be classified as scholarship that is focused primarily on agency. We cannot change one term of Burke's pentad without affecting the others in kind, certainly; nevertheless, the introduction of new information technologies into our classrooms are, first and foremost, represented as new *ways of doing* education. E-mail and chat rooms provided new means of connecting with others, hypertext promised the rise of a more active reader (i.e., one with more agency), and the vision of the Web as a global information network suggested that students and researchers would have near-infinite resources at their fingertips. Much of the criticism that has followed first-wave adoption and optimism, in fact, has focused indirectly on our tendency to view technologies as pure agency without reflecting upon the inevitable effects they might have on our status as agents, the scenes we occupy, and so on. I raise this issue of agency because I think that we might learn from Burke's regret and nuance our own considerations of agency more carefully. There is a difference between the matter of agency (e.g., tools, methodologies) and the manner of agency (e.g., attitudes, styles), a difference that is often elided in our scholarship.

Nowhere is this distinction more evident than in the way we have approached linking. The link is one of the minimally sufficient conditions for hypertexts and Web pages and as such, has received a great deal of critical attention. For the first decade or so of hypertext theory, according to Katherine Hayles, the link tended to loom larger than hypertext's other characteristics (27). But the attention that we have paid to linking is marked by the confusion between tool and attitude; too frequently, critics substitute their meditations on the manner of linking for careful interrogation of the matter or materiality of links. The failed promises of hypertext theory provide perhaps the most notorious example of this substitution. The presence of multiple reading paths allows readers to adopt an *attitude* toward the text that is more active, but in hypertext theory, this often translates into ill-advised claims that the presence of links themselves necessarily produces such readers—claims that are still replicated in scholarship today. Peter Lurie, for example, in his essay "Why the Web Will Win the Culture Wars for the Left: Deconstructing Hyperlinks," argues that the very structure of the Web will produce a particular cultural outcome: "The architecture of the web, and the way users navigate it, closely resembles theories about the authority and coherence of texts that liberal deconstructionist critics have offered for thirty years. . . . Anyone who has spent a lot of time online, particularly the very young, will find themselves thinking about content—articles, texts, pictures—in ways that would be familiar to any deconstructionist critic. And a community of citizens who think like Jacques Derrida will not be a particularly conservative one."[1]

Lurie continues, arguing that these basic technical tools are similar to deconstructionist analytical tools, but he fails to mention that, in Derrida's case particularly, those deconstructionist analytical tools are simply the analytical tools provided by the philosophical tradition. Derrida breaks from that tradition in the manner he employs those tools. Lurie repeats many of the problems with first-wave hypertext criticism, not by perceiving a similarity between the Web and poststructuralist theories but by posing their relationship as homology.

The problems with that relationship become clearer when we explicate the difference between the matter and the manner, particularly in the context of the link. There is a fair range of scholarship on linking, and a focus on manner or attitude pervades this work. In many cases, what this scholarship either elides or outright lacks is an awareness of the materiality of the link. I do not mean to suggest that we turn to this materiality *instead of* manner but rather that our discussions of link styles will continue to suffer until we begin to account for the nuanced relationship between manner and matter. Lurie is correct to identify hyperlinks as a crucial component of our new media grammar, but an effective rhetoric of linking demands that we consider links as both tools and styles.

Manner

The most exhaustive treatment of links remains George P. Landow's *Hypertext*, and considerable space in that book is devoted to the subject. Landow's most concentrated discussion of linking takes place within the larger context of orientation practices; links are suggested as one of the central features for helping readers orient themselves. Landow's understanding of links is an explicitly spatial one: "Since hypertext and hypermedia are chiefly defined by the link, a writing device that offers potential changes of direction, the rhetoric and stylistics of this new information technology generally involve such change—potential or actual change of place, relation, or direction" (124).

The notion that links represent a change of direction persists across Landow's entire discussion of links; he proposes that we think in terms of a rhetoric of arrival and a rhetoric of departure to account for the fact that creators must decide what readers need to know at either end of a hypermedia link in order to make use of what they find there (124).

Landow's attention to distance would seem to imply a corresponding focus on the potentials of a spatialized text, but surprisingly, Landow ends up severely restricting this metaphor. He writes, "Because hypertext linking takes relatively the same amount of time to traverse, all linked texts are experienced as lying the same 'distance' from the point of departure" (125). This claim is questionable at best, but most importantly for my own purposes, it reveals the degree to which the materiality of linking is invisible in Landow's work.[2] If all nodes are equidistant and traversing a link is relatively instantaneous, then Landow's analysis effectively erases the concept of distance altogether.[3] Despite his careful attention to the variety of hypertext platforms available, Landow reduces the link itself to an invisible connection between nodes, a nearly immaterial pause between pages that itself has no appreciable effect on the nodes it links together. And this characterization of links holds true across all the platforms that Landow examines.

As a result, the rhetorics of departure and arrival have less to do with links than they do with the nodes that are linked. Both rhetorics are presented in terms of utility and efficiency; according to Landow, the very existence of links conditions the reader to expect purposeful, important relationships between linked materials (125). Landow's example for a rhetoric of departure is both obvious and mundane: "For example, when an essay on Graham Swift or Salman Rushdie uses phrases like World War I or self-reflexive narrators, readers will know that links attached to those phrases will lead to material on those subjects" (152).

Landow's discussion of the rhetoric of arrival is similarly mundane. He lobbies for functions such as those in Storyspace and InterMedia that provide highlighted sections to indicate precisely where links are intended to arrive. In both cases, links themselves are functionally invisible; what happens at either end of a hypermedia link is of far more importance for Landow. His work subsumes links themselves into the departures and arrivals that take place in the nodes, and this, combined with his penchant for minimizing the material differences among hypertext authoring platforms, results in a definition of hypertext in which linking, quite literally, doesn't matter. In Landow's account, linking is clicking, an interactivity defined in mechanistic, yet transcendent terms, according to Michelle Kendrick, and one that trivializes hypertext. Materially, links are invisible for Landow, and because they only function to signify the purposeful relationship of "more" or "next," there is no real style to the hyperlink, either.

This position is not universal to all hypertext critics, however. Nicholas C. Burbules, in "Rhetorics of the Web: Hyperreading and Critical Literacy," is critical of perspectives, such as Landow's, that render links trivial. He explains, "My hope is to invert the order of how we normally think about links and information points, nodes, or texts: usually we see the points as primary, and the links as mere connectives; here I suggest that we concentrate more on links—as associative relations that change, redefine, and enhance or restrict access to the information they comprise" (103).

Burbules, like Landow, focuses on rhetoric; unlike Landow, however, Burbules's essay focuses on links as rhetorical devices. Burbules introduces what he describes as a menagerie of links, borrowing terminology from rhetorical tropes to characterize the various types of connections that can be deployed via links.[4] As he explains, Burbules inverts the relationship between text and link that Landow proposes, asking links to bear a rhetorical burden that Landow places solely within the texts that are linked. If, for Landow, both the manner and matter of links are invisible (and invariable), Burbules rescues the notion of manner and of various link styles.

To his credit, Burbules is careful to point out that there is nothing about the form of such materials that *ensures* more perspicuous readings or new ways of organizing information (107). In other words, there is no guarantee that the identification of various link styles, or even the conscious use of them, will translate into user experience. He also notes that, in addition to semic relations, links establish pathways of possible movement within the Web space (Burbules105). With each of these observations, Burbules hints at the possibility that there is something beyond the manner or style of linking that may have an impact on our experiences of linked texts. However, he doesn't carry either observation far enough; his focus is squarely on the manner of linking. In fact, he begins his discussion of linking by effectively abandoning any attempt to come to terms with the materiality of links: "[A]lthough all links in a hypertext work in the same way, involve the same act (clicking on a highlighted word or icon), and end with much the same result (a new screen appearing), all links are not the same, and do not imply the same type of semic relation" (104). This explanation of the material behavior of links is insufficient, particularly when compared to the detail that Burbules provides to explain the types of semic relations. Burbules is critical of analysis that renders the link-event invisible, but his essay makes the link visible even as the event vanishes. If all hypertexts work in the same way, then there is no point to including the material event in our understanding of the link.

Jeff Parker's "A Poetics of the Link" provides an important counterpoint to Burbules's account. Parker's goal in this essay is similar to Burbules's; he is interested in providing a taxonomy of links, although his is based less upon an overarching system (such as rhetorical tropes) than it is grounded in his own experience as a hypertext writer. Parker offers as an example a hypertext story he has written, "A Long Wild Smile," and describes a number of different link types based on that composition.[5] Although Parker himself makes little of this, it is worth examining his description of the hypertext: "The main structural layout of this piece is two texts side by side, both in the first person. . . . For the most part each node has at least two links in it, one that links to its partner narrative, another that links within its own narrative" (par. 14).

Like Burbules, Parker is concerned more with the manner or style of his links than with the material event of linking, but this description raises an interesting distinction. In Parker's text, links do not work the same way. Some of them (links within its own narrative) work the way we typically expect of Web-based links; they cause the node with the link to vanish and a new node takes its place. But others, those that link to the partner narrative, bring up a new node without replacing the node that contains the link.

This is not an earth-shattering distinction; rather, this difference has been an affordance of Web-based hypertext for as long as browsers have been capable of reading frames. But it raises an interesting issue with respect to the materiality of linking. The two halves of Parker's text are written in intentionally contrasting styles, but that difference is underscored by the material effect that the self and partner link types provide. A partner link allows the reader the opportunity to see both the link and the resulting node side-by-side, which would encourage (though not ensure) the consideration or recognition of the semic relations that both Parker and Burbules advocate. More to the point, however, Parker's discussion unwittingly acknowledges a distinction that is at least as important as the menageries or taxonomies of link styles. The link-event, the link's material consequences for user experience, remains undertheorized in our explorations of new media.

From a certain perspective, this is understandable. Despite popular claims to the contrary, we are still firmly ensconced in print culture, in which the material navigation of texts is, with very few exceptions, limited in scope to the turning of pages. Landow, Burbules, and Parker represent a small fraction of the critics who have examined links not as matter but as manner, as tools for stylistic variation rather than formative elements of the technology itself. Such examinations are rooted in traditions (such as rhetoric or poetics) that have been tied to print, a medium in which material invisibility is the ideal. Connective or transitional language in print is indeed an issue of style, manner, and attitude, but insofar as linking usurps that function in new media, we must consider in more detail the degree to which the materiality of the link-event shapes our experiences of such texts.

Matter

A second element of Parker's essay that is important for our purposes here is his invocation of Espen Aarseth's work on cybertext. Unlike much of the hypertext criticism that precedes (and even follows) Aarseth's *Cybertext*, Aarseth's work pays close attention to the materiality of new media, so much so that he decries attempts to understand such work that fails to attend to materiality. Parker cites Aarseth's discussion of ergodic textuality, those works that require nontrivial effort on the part of a reader for navigational purposes. Parker takes Aarseth to mean that the activation of a hypertext link, in his definition, is nontrivial, although this is debatable at a time when large portions of our society are comfortable clicking on Web links. This is another place where Parker elects to understand linking in terms of manner rather than matter, for the nontrivial effort he goes on to describe takes place at the level of semic relation.

However, Aarseth is careful not to restrict his understanding of "ergodic" in this way. Many of Aarseth's examples, in fact, such as MOOs (multiuser domains, object-oriented) and video games, do not mean anything in the same sense as a short story means something. In Aarseth's terms, the interpretive role played by a short story reader is not a necessary component of all cybertexts. Nontrivial effort with respect to ergodic texts may involve anything from the shuffling and dealing out of a tarot deck to the manipulation of an avatar in a video game; the range of activity certainly includes reading as we traditionally conceive of it, but reading and interpretation do not exhaust Aarseth's understanding of ergodic textuality. Analogously, it is shortsighted to imagine that the blue, underlined word or phrase that disappears and summons a new page on the Web is our only conceivable means of linking.

In addition to the work of Aarseth, Hayles, and those few others who have begun to attend to the materiality of new media, we might also think of usability studies as a site where such materiality is being investigated. Links are but one of the elements that usability experts might choose to examine, but their overriding concern is with the material elements of technology: physical accessibility, information architecture as it impacts users' ability to retrieve data, and page layout as it aids or hinders the efficient access of information. In each study, usability measures a page or node according to factors such as ease of use, time spent on task, and so on—the material signs of user access.

In some ways, cybertext and usability are inversely related. If Aarseth is specifically concerned with nontrivial effort, the work of a usability expert such as Jakob Nielsen is devoted to rendering user effort trivial. Aarseth is interested in charting the variety of cybertexts, whereas usability is more frequently concerned with minimizing that variety, or measuring technology against a standard driven by transparent efficiency. Aarseth is not an advocate of variety for variety's sake, certainly, but nevertheless we might see these two emerging fields of study as poles between which a spectrum of theories about the materiality of technology would lie. I would locate this chapter somewhere along that spectrum, certainly. Its central claim—that we should attend to the materiality of linking practices as carefully as we examine their manners—could not be reconciled with orthodox usability theory; there is value, I would argue, in exploring the expressive potential of materially distinct links—value that would be short-lived if we were simply concerned with efficiency.

That concern, implicit in so much first-wave hypertext criticism, comes not only from the fact that we are embedded in print culture, however. It is also the result of our attempts to place new media within the history of print literacy. From theories of poststructuralist textuality to pronouncements like Robert Coover's about the end of books, early hypertext critics located themselves within a print or literary tradition, and the triviality of linking is a product of that location.[6] Aarseth describes this as media chauvinism, and a portion of his book is spent refuting this tendency in hypertext criticism.

Perhaps the most sustained refutation of this particular historicization comes from Lev Manovich, whose *The Language of New Media* begins not from language per se but from cinema. A hundred years after cinema's birth, Manovich explains, cinematic ways of seeing the world, of structuring time, of narrating a story, and of linking one experience to the next have become the basic means by which computer users access and interact with all cultural data (78–79). Manovich's demonstration of this claim comprises the majority of his book. For our purposes, though, it is useful to consider where first-wave hypertext criticism matches up with Manovich's discussion. The kind of linking advocated by Landow is only capable of raw connection or juxtaposition; in this, it bears some resemblance to the earliest theorization of cinematic montage. Again, though, Landow's model for the link simply suggests the possibility of editing. Burbules, in attempting to categorize various semic relationships through links, introduces a Kuleshov Effect into the practice of linking. Kuleshov was the Russian director who demonstrated that juxtaposing various images with an actor's face would lead audiences to attribute different (and potentially conflicting) emotions to the expression. While Burbules names these various relations (that correspond to the emotions that Kuleshov evoked), Parker goes one step further in placing the juxtaposed nodes on the same screen. With Parker's hypertext, the montage becomes more self-conscious. But none of the three go beyond manner in their discussions; montage or juxtaposition may be capable of producing a broad range of relationships, but in the end, it is a single type of connection.

My point here is not to denigrate these critics by suggesting that they are somehow behind the times; rather, I share with Manovich the belief that we have yet to draw on our full range of cultural resources when it comes to thinking about new media practice. Specifically, I would argue that we need to move beyond print in our conceptions of linking. Manovich's emphasis on cinema can be particularly productive in this regard. We are so familiar with cinematic technique that even nonspecialists can probably recognize and define terms like fade, jump cut, montage, pan, and zoom. And when Gunther Kress and Theo van Leeuwen explain that visual close-ups connote an intimate or personal relationship with the person being depicted, most of us would feel that their explanation is simply confirmation of what we understand intuitively. It is easy to imagine a scene in a movie that opens with a wide-angle shot of a crowd or group of people and then the camera zooms in for a close-up on a particular face, perhaps the face of the character central to that scene. As an audience, we have moved from an impersonal understanding of that character's context to a shot where we might expect more personal information or action on behalf of that character.

The connection that we have yet to make, at least in terms of theorizing new media, is that it is possible to read each of those shots as nodes and the zoom as the link between them. And the effect achieved by a zoom differs from what would be signified by a simple cut; a shot of the crowd juxtaposed with a facial close-up would probably suggest that the character is watching the crowd, not that he or she is part of it. These disparate meanings, in the context of this essay, are different manners of presentation, manners that rest upon the material base of the camera or the cinematography. Burbules and Parker both demonstrate a desire to expand the practice of linking beyond the trivial click, but that desire must accompany a corresponding interest in developing links as material events. The print tradition, driven as it is by standardization and technological invisibility, is not going to be of much help.

Matter and Manner

What does it mean to think of links as material events? The most common form of link, a form with which any regular user of the Web will be familiar, is the word or phrase on a Web site coded with the ubiquitous <a href=>. Such links are most often blue (or purple if they link to visited nodes) and underlined, which sets them apart from standard text. The underline feature of links is so ingrained by this time that it is exceedingly rare (and ill-advised) to use underlining for any other purpose on the Web. Typically, we access a Web link by bringing our cursor over the word so highlighted and clicking on it; when we do so, the node or page in our browser window disappears and the link's target page appears. There are some exceptions to this experience, but they do little to complicate the click. One comes from the example in Parker's essay, when a link in one frame summons content to another frame. Another obvious exception is the link that opens a new browser window rather than replacing the content of the one that holds the link. Exceptions like these are more variation than anything else, and like print, the ubiquity and persistence of uniform linking practices on the Web has the effect of naturalizing it, rendering it invisible, and reducing it to the click of the mouse.

I want to suggest another exception to the typical user experience of the Web link, one that I make use of in a multimedia essay published in 2001 in *Enculturation*: "Perspective: Notes Toward the Remediation of Style" (Brooke). Composed with Macromedia Flash, this essay is not strictly a hypertext but rather a predominantly linear series of nodes (called scenes in Flash) that function in part like Web pages. That is, each scene contains several pauses—moments that require the user to access a link to start the essay in motion again. Links are indicated differentially; users are instructed to click on words that differ in color from the rest of the paragraph in which they appear. In this way, "Perspective" is much like a Web-based essay, and in fact, it draws on this similarity to minimize any confusion on the part of users.

However, Flash is not a platform for designing static Web pages, owing much of its internal vocabulary to cinema; in Flash, one composes movies that are made up of scenes, themselves further subdivided into individual frames and displayed (normally) at twelve frames per second. The motion that Flash permits prompted me to rethink the materiality (as well as the style) of the links I built into the essay. Pages rarely vanish; instead, elements of the essay (e.g., text blocks, graphics) move on and off the screen in a specifically choreographed fashion. One of the central features of Flash is that it allows a writer to work with many layers of graphics and text; individual elements of a scene may be altered independently of the others. Unlike standard Web links, which must normally disappear as part of the page is being replaced, a link implemented in Flash need not do so. At several points in "Perspective," for example, clicking on a link will render a text block visible without replacing the original text. In Figure 1, the phrase "reading praxis" appears

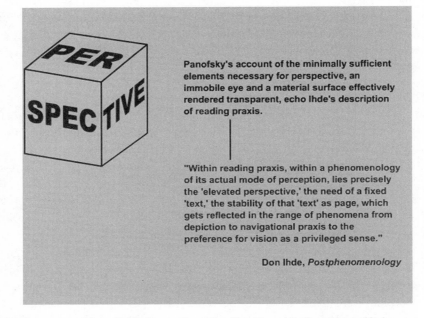

Figure 1.   Screen capture from the Flash essay *Perspective: Notes Toward the Remediation of Style*

Figure 2. Screen captures from *Perspective: Notes Toward the Remediation of Style*

as a link to access, and the line and the lower paragraph emerge on screen once that link is accessed.

But this is not substantially different from the frame-to-frame linking that Parker describes, and so I want to turn to a later scene from "Perspective," one where something quite different happens with a particular link (see Figure 2).

On this screen in Figure 2, a transition is effected from a discussion of the hermeneutic, interpretive positioning of book and hypertext readers to another text block that compares it to the spatial, perceptual positioning in video games. When a reader clicks on the word "positioning," the graphic and much of the text are replaced on the screen, but the linked word itself is not. Instead, it drifts upward to occupy a new position itself, this time in the new sentence. The text here is making an explicit argument for the productive ambiguity of the word "position," but that argument is implicitly underscored by the reader's experience of the word itself, which plays a role in two different passages. There are several places where similar wordplay takes place; the term "perspective" itself denotes both the visual, perceptual monologic that defined art for centuries and the partial, contingent opinion held by a person in a given context.

Could such effects be achieved on Web pages? Perhaps, through the careful use of dynamic layers and precision layout. But it is worth noting that many such effects occur in "Perspective" not through conscious planning but rather as a result of my exploration of the platform. It is also significant that much of that essay is composed and designed with predominantly basic features of Flash; one would not need to be more than a basic user to duplicate much of what "Perspective" accomplishes. In other words, even an initial familiarity with tools like Flash allow us to conceive of links in a more material fashion than most hypertext theory acknowledges.

The purpose of this chapter, however, is not to advocate for a particular platform but rather to advocate for a particular practice. Tools such as Flash enable us to push beyond the models for new media expression offered to us by Web-based HTML. We should not stop asking ourselves, as readers or writers, what a given link means or what manner of link it is, but those sorts of questions should be supplemented with what I have come to think of as an even more fundamental question. Put simply, we should begin asking ourselves what *can happen* when we access links. As new media grow in popularity, there is no need to restrict ourselves to the blue underline any more than we restrict ourselves to the page.

Lifting such restrictions will allow us to nuance our understandings of new media, much as the separation of agency into tool and attitude did for Burke. The point is not to separate them completely—any more than we can separate form and content more generally—but rather to examine and consider those ways that an expanded materiality of links might prove to be a more effective counterpart to the work that has already been done on the manner or styles of such connections. If new media are to go beyond the mechanical paucity of the click, then as writers we need to rethink the vanishing, underlined word or phrase as our only option for connecting one node to another. We need to draw on those cultural traditions that have already done much of that rethinking and incorporate their insights into our own.

Notes

1.   Michelle Kendrick is particularly critical of claims like Lurie's: "Hypertext may well enable very different kinds of writing and different kinds of reading; it was not my purpose here to deny that. My claim is that electronic media are *not* in themselves 'revolutionary': to understand critically the myriad effects of such new technologies one needs a sophisticated, theoretically informed understanding of 'history,' 'technology,' and 'theory.' In exploring the potential, over the next decade, educators and theorists must be careful not to do what many educators and theorists have done over the last decade: turn reading into consumerist browsing; value technology, in and of itself, as pro forma endowing students with critical thinking skills; valorize 'click' choice over critiques of systems of choice; and demean acts of print-reading and the activity of comprehension in favor of a mechanistic sense of interactivity" (249–50).

2.   Landow explains that "the goal is always potentially but one jump or link away," hence all other nodes are equidistant from the node that a reader currently occupies. But this is an impoverished account of the readerly experience. Technically, the sixth page I visit in a hypertext may indeed be a link away from the front page, but I will *experience* it as further away from that front than the third page I visit. We do not *experience* hypertexts per se; rather, we can only experience our readings of a hypertext, and the order of our reading will translate into our experience of such "distance." This claim also ignores the formative power that hierarchies of folders and files (the infrastructure of many Web sites) holds for readers.

3.   What delay does exist is experienced (and theorized) as a property of the page (loading time) rather than the link itself.

4.   Burbules separates link types into metaphor, metonymy, synecdoche, hyperbole, antistasis, identity, sequence, and cause-and-effect.

5.   Parker separates links into two broad classifications: functional (blatant, filler, random) and literary (emotive, lateral, complicating, temporal, portal).

6.   In many cases, linking is likened to the process by which readers of a book trace a text's footnotes. This laudatory analogy often serves to demonstrate that linking is even less intrusive or labor-intensive than the turning of a page.

Works Cited

Aarseth, Espen J. *Cybertext: Perspectives on Ergodic Literature*. Baltimore: Johns Hopkins University Press, 1997.

Brooke, Collin Gifford. "Perspective: Notes Toward the Remediation of Style." *Enculturation: Special Multi-journal Issue on Electronic Publication* 4.1 (Spring 2002). http://enculturation.gmu.edu/4_1/style.

Burbules, Nicholas C. "Rhetorics of the Web: Hyperreading and Critical Literacy." *Page to Screen: Taking Literacy into the Electronic Era*. Ed. Ilana Snyder. London: Routledge, 1998. 102–22.

Burke, Kenneth. *Dramatism and Development*. Barre, Mass.: Clark University Press with Barre Publishers, 1972.

Coover, Robert. "The End of Books?" *New York Times Book Review* 21 June 1992: 11, 23–25.

Hayles, N. Katherine. *Writing Machines*. Cambridge, Mass.: MIT Press, 2002.

Kendrick, Michelle. "Interactive Technology and the Remediation of the Subject of Writing." *Configurations* 9.2 (Spring 2001): 231–51.

Kress, Gunther, and Theo van Leeuwen. *Reading Images: The Grammar of Visual Design*. New York: Routledge, 1996.

Landow, George P. *Hypertext 2.0: The Convergence of Contemporary Literary Theory and Technology*. Baltimore and London: Johns Hopkins University Press, 1997.

Lurie, Peter. "Why the Web Will Win the Culture Wars for the Left: Deconstructing Hyperlinks." *CTHEORY*. 15 Apr. 2003 http://www.ctheory.net/text_file.asp?pick=380.

Lyotard, Jean-François. *The Differend: Phrases in Dispute*. Minneapolis: University of Minnesota Press, 1991.

Manovich, Lev. *The Language of New Media*. Cambridge, Mass.: MIT Press, 2001.

Parker, Jeff. "A Poetics of the Link." *Electronic Book Review* 12 (Fall 2001). http://www.altx.com/ebr/ebr12/park/park.htm.

8

ScriptedWriting() { Exploring Generative Dimensions of Writing in Flash ActionScript

David M. Rieder

> This slips into that; writing slips into programming; programming slips into writing; depending on the view we take of "writing," "programming" is a subset of "writing" or is distinguishable from "writing" . . . the answer is yes and no.
>
> —Jim Andrews, "[empyre] Welcome Jim Andrews Re: Electronic Poetry"

If ((Code == Text) || (Code != Text)) {

At what point, and with what disciplinary values in mind, should we distinguish (natural language) text from (artificial) code? In the epigraph, echoing a small but growing number of other commentators across the humanities, Jim Andrews observes that the line between the two linguistic species is variable. Nonetheless, there is an inertial tendency in the humanities that continues to reinforce the value of linear or alphabetic writing over the hybrid "postalphabetisms" of programmed code writing. In other words, humanities scholarship is disposed to focus on phonetic thought and communication rather than decentering this species of communication for a richer approach that values the emerging litany of postalphabetic dimensions of writing (such as the spatial, visual, kinetic, and algorithmic). While theorists and commentators of digital writing may have observed the "slip" to which Andrews refers, mainstream scholarship is persistently focused on linear, alphabetic, natural language texts.

One of the ways in which this phonetic-linear inertia is being challenged, however, is through calls for a renewed emphasis on production or invention over the traditional concern for critique and analysis. As techniques and technologies for the production of texts become increasingly available and as a generation of influential voices in the humanities focuses on methods of invention over representational critique, there is a rising trend in current scholarship to blur the disciplinary lines that conventionally separate text from code. For example, in a recent article titled "Deeper into the Machine: Learning to Speak Digital," N. Katherine Hayles urges her audience to develop a hybrid discourse to bridge the divide between text and code. With an interest in recent electronic literature known as *codework*, she urges her readers to develop a critical vocabulary for the expanding range of emerging semiotic activities. In the following quote, she defines some of the characteristics of this vocabulary in terms of interplay and integration: "This new critical vocabulary will recognize the interplay of natural language with machine code; it will not stay only on the screen but will consider as well the processes generating that surface. . . . [I]t will toss aside the presupposition that the work of creation is separate from the work of production and evaluate the work's quality from an integrated perspective that sees creation and production as inextricably entwined" (373).

Hayles's call to think beyond the conventions of any one field and to recognize the extent to which creation and production are "inextricably entwined" is one of a number of recent attempts to develop a crossdisciplinary or interdisciplinary perspective in digital writing studies—one that recognizes the ways in which code and text are inextricably linked. As I will illustrate in what follows, the concept of "interplay" is an important step away from the phonocentric inertia targeted in this essay but too broad of a description for the manifold ways in which text and code are coordinated in current projects. In application environments like Flash (in which on-screen textual experiments are scripted "off screen" in coordination with the ActionScript programming language), there is an opportunity to catalogue the new interlinguistic dynamics that programmer-writers are developing. Like a study of rhetorical schemes and tropes, writing theorists and practitioners will have to develop studies of the manifold ways in which text and code are scripted in concert in digital environments. Following a section in which I highlight a few of the ways in which the question of interplay is called into question by in-depth studies of text and code, I present brief studies of two Flash-based projects published in *Born Magazine*. Both projects move well beyond the use of tweens, and button-based events to develop generative writing environments with the ActionScript programming language.

ActionScript is a high-level, object-based, proprietary language that is a part of Adobe/Macromedia's Flash application. I choose this language primarily for pedagogical reasons. First, as a high-level language, ActionScript has more in common with human languages than with assembly languages, which are closer to the language of the machine. With this aspect in mind, high-level languages such as ActionScript are a useful environment in which to practice writing, because we can delve more "deeply" into the machine while still retaining a connection to the human-natural language arts on which our fields are based.

Like a number of other well-known high-level languages (e.g., FORTRAN, Pascal, C, Perl, PHP, and Javascript), ActionScript is dependent on the computing environment in which it is used. From a pedagogical standpoint, a high-level language like ActionScript is easier to read and write, and since it resembles human languages (primarily English), its use is apropos. Furthermore, as an object-based language, ActionScript contains a predefined set of prefabricated objects that can be utilized with relative ease to explore increasingly sophisticated dimensions of electronic textuality.

A generative writing environment is a particular kind of digital writing environment that breaks free of the implied instrumentalism of writing as an extension of voice. It creates a screenic environment dynamically adjustable to user input or to the aleatoric processes of the computing environment. The first of the two Flash projects that I study is called "Crushed under Sodium Lights," by Dave Bevan and Brendan Dawes. This project enables its user to interact with poetic lines of text along a number of interrelated dimensions of writerly expression that include the spatial, visual, kinetic, and temporal. In the second, "Stark, North of Gainsboro," David Deraedt and Qubo Gas redistribute a poem by Joan Houlihan along spatial and kinetic dimensions of expression using a function that randomizes some of the movements across the screen. As the movie plays, words break free of their verses and letters from the words to which they were bound, Houlihan's poem transformed into a compelling example of writing "after the voice."

In both projects, the ActionScript programming is an essential part of the project. The use of modern programming concepts like loops and conditionals, as well as object-based programming techniques related to objects and methods, are inextricably linked to the interactive and aleatoric dimensions of these projects. For this reason, a reading of the writing scripted on-screen does not adequately dramatize the extent to which writing has become—and should be studied as—a differential process of coordinations between text *and* code.

Textual Challenges: From Transparency to Opacity

In his essay, "Surf-Sample-Manipulate," Mark Amerika describes his method of sampling and manipulating the source code from HTML pages on the World Wide Web to create his hypertext project, "GRAMMATRON." Linking his method of composition to the modern art of collage in cubist painting and Dadaism, Amerika blurs the lines between content (i.e., text) and source code: "The great thing about the Net is that if you see something you like, whether that be 'content' or 'source code,' many times you can just download the entire document and manipulate it to your needs. In fact, it wouldn't be entirely suspect to suggest that 'content' and 'source code' are one and the same thing, since as far as the Web goes, one cannot simply exist without the other" (par. 9). Importantly, "GRAMMATRON" was developed in the mid to late nineties, when innovative Web content based on static page design was still an acceptable norm. Except for graphical and sound files (and the occasional Java applet), Web-based content and source code were often found together in the same page. In other words, the Web was experienced as a "text-based" programming environment, a phrase that expresses the extent to which the line between text and code were blurred.

Borrowing Sherry Turkle's distinction between transparent and opaque computing interfaces, I argue that the Web depicted in "Surf-Sample-Manipulate" is one in which code and text are transparent (i.e., are coexistent).[1] Today, the Web is increasingly opaque, supporting the idea that code and text come from different linguistic regimes. The code that an artist like Amerika might access in a Web page is often written dynamically by server-side scripts. Alternatively, the code is little more than a framework for the publication of a compiled, third-party application. Compared to "first generation" hypertexts like Amerika's "GRAMMATRON," "second generation" cybertexts and Flash-based works require a new level of bilingualism, contributing to arguments such as that developed in John Cayley's essay "The Code is not the Text (unless it is the Text)."

Cayley criticizes what he characterizes as a "utopian" gesture in some recent theory and artistic practices that celebrates the commensurability of the two linguistic species. For example, in projects by Mez, natural and artificial languages are "creolized" in order to explore what Rita Raley describes as "hybridized, electronic English, a language not simply suggestive of digital and network culture but a language of the computer and computing processes" ("Interferences" par. 15). Cayley's issue with the criticism that unilaterally accepts the implicit linguistic proposition developed in Mez's work is that this creolized language is meant for the screen, not for the computer and its internal processes: "The utopia of codework recognizes that the symbols of the surface language (what you read from your screen) are the 'same' (and are all ultimately reducible to digital encoding) as the symbols of the programming languages which store, manipulate, and display the text you read" (par. 4). Mez's hybrid language, *m[ez]ang.elle*, which utilizes some of the syntax of the Perl language, is not operational. Her language does not address both humans *and* machines. For Cayley, code and text are expressive of fundamentally different operational regimes.

In the following quote, Cayley explains why text and code are separate and distinct species of writing: "[B]ecause code has its own structures, vocabularies, and syntaxes; because [code] functions, typically, without being observed, perhaps even as a representative of secret workings, interiority, hidden process; because there are divisions and distinctions between what the code is and does, and what the language of the interface text is and does, and so on" (par. 5). Highlighting some of Cayley's own vocabulary, operational code is "secret," internal, and "hidden," adjectives that are defined in opposition to the visible, external textuality "on the screen." Perhaps for this very reason, Cayley argues that text and code require different reading strategies, which will eventually lead to "better critical understanding of more complex ways to read and write which are commensurate with the practices of literal art in programmable media" (par. 5).

Cayley's argument is problematized by humanities-oriented computer scientists like Donald Knuth and Geoff Cox. In *Literate Programming*, Knuth argues for a new "attitude" in computer programming. He writes, "Instead of imagining that our main task is to instruct a computer what to do, let us concentrate rather on explaining to human beings what we want a computer to do" (14). Whereas Cayley asked that we distinguish text from code, Knuth wants to blur the distinction. Along the same lines, while Amerika's implicit call for this "attitude" is based on his experience with "first generation" Web work, Knuth is working with the kinds of "hidden" or "secret" depths of code about which Cayley is writing.

Knuth goes on to equate computer programs with literature and to redescribe the programmer as an essayist. Knuth draws out the analogy in terms of human comprehensibility: "Such an author, with thesaurus in hand, chooses the names of variables carefully and explains what each variable means. He or she strives for a program that is comprehensible because its concepts have been introduced in an order that is best for human understanding, using a mixture of formal and informal methods that nicely reinforce each other" (99). So, Knuth is arguing for a blurring of the lines between code and text, even while he is working with "hidden" or compiled code—code that would otherwise support the idea that code and text represent divergent strategies of writing and reading.

In another related argument, in "The Aesthetics of Generative Code," Geoff Cox, Alex McLean, and Adrian Ward evaluate the aesthetic dimensions of code, linking its logic and syntax to poetry in ways that would also contradict Cayley. As if in reply to Cayley, Cox, McLean, and Ward write, "To separate code and the resultant actions would simply limit the aesthetic experience, and ultimately limit the study of these forms—as a form of criticism—and what in this context might better be called a 'poetics' of generative code" (par. 24).

For Cox, McLean, and Ward, the textualities of computer code and poetic text are comparable in their relations with the performative or generative dimensions of their composition.[2] They write, "code works like poetry in that it plays with structures of language itself, as well as our corresponding perceptions. In this sense, poetry might be seen to be generative in that it is always in the process of becoming" (par. 16). As they go on to stipulate, computer code is not a static entity. Code is inextricably linked to the "internal structure that the computer is executing, expressing ideas, logic, and decisions that operate as an extension of the author's intentions" (par. 19). Like poetry, code relies on a larger context of performative possibilities.

As they explore this issue, Cox, McLean, and Ward write, "This synthesis suggests that written and spoken forms work together to form a language that we appreciate as poetry. But does code work in the same way?" (par. 9). In their initial example, an award-winning Perl poem by Angie Winterbottom, they compare the original text to the coded translation:

If light were dark and dark were light
The moon a black hole in the blaze of night
A raven's wing as bright as tin
Then you, my love, would be darker than sin.

```
if ((light eq dark) && (dark eq light)
&& ($blaze_of_night {moon} == black_hole)
&& ($ravens_wing {bright} == $tin {bright})) {
my $love = $you = $sin {darkness} +1
};
```

Comparing the two versions, several initial comparisons stand out. The "if" in the original text is paralleled by the "if" statement in the code. The past tense of the verb "to be" is represented by the equivalent operators in Perl, namely "eq" and "==." Finally, the conjunction "and" is paralleled in the code by the operator "&&." As we analyze the two poems more closely, another two levels of comparison stand out. First, Winterbottom has re-presented some of the nouns and noun phrases in the original text as variables in the coded version. For instance, "blaze of night" and "tin" have a "$" appended to them, which signifies their transformation into Perl code. Moreover, the use of parentheses and braces in the coded version expresses logical relationships between the ideas and images in the original text. In an attempt to "speak digitally," for which Hayles calls, a reading of the coded version would lead us to appreciate the visual dimensions of the meaning expressed by the nonphonetic or postphonetic parentheses, curly braces, and variables. Like Mez's Perl-inspired *m[ez]ang.elle*, text and code are combined in ways that seem to reflect

Amerika's experience of Web-based writing. The distinctions between code and text are one and the same thing.

According to Cox, McLean, and Ward, Winterbottom's Perl poem is a disappointment in that it represents little more than a transliteration of the original text-based poem into the Perl language: "Poetry at the point of its execution (reading and hearing), produces meaning in multitudinous ways, and can be performed with endless variations of stress, pronunciation, tempo, and style" (par. 11). Winterbottom's Perl poem is not meant to be operational. As Raley explains of Mez's codework, the "code-language" is not a functional program. It is a "static art artifact." Raley writes, "[H]er codework is technically non-referential and technically non-performative" ("Interferences" par. 26). Like Mez, Winterbottom does not tap into the generative dimensions of computer programming, which Cox, McLean, and Ward consider the prime moment at which text (poetry) and code relate. They write, "Crucial to generative media is that data is actually changed as the code runs" (par. 14). When the various routines in a program are executed—generating a performative space of operative meaning—code and text, programming and poetry, operate on the same level of linguistic meaning. Like poetry, code must be executed for us to fully appreciate its aesthetic value—as well as the relations among the two forms of language use. An example of an executable poem is shown in Figure 1.

On the "surface," Ted Warnell's visual poem, "LASCAUX.SYMBOL.IC," is an example of codework in which code-as-text can be operational. In the poem, the following short, one verse JavaScript is presented as the text:

```
1    <SCRIPT language="Javascript">
2    <!—
3    var d = new Date();
4    var t = d.getTime();
5    var x = ( t % 9 ) + 1;
6    document.write( "0" + x + "<br>" );
7    document.write ( "[0" + x + "]" );
```

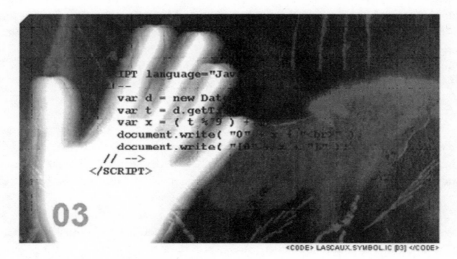

<CODE> LASCAUX.SYMBOL.IC [03] </CODE>

Figure 1.   Ted Warnell, "LASCAUX.SYMBOL.IC"

```
8    // ->
9    </SCRIPT>
```

Although the code-as-text is nonoperational, when it was copied and pasted into an HTML page, it generated the following two lines of verse at exactly 5:58.12 p.m., eastern standard time, Saturday, April 29, 2006:

```
03
[03]
```

The code uses the built-in Date() object and .getTime() method to read the exact date from a user's browser, in milliseconds, since the strike of midnight on January 1, 1970, Greenwich mean time. (At the day and time mentioned above, the number was 1146347892317). The modulus operator is then used to divide the number by nine and determine its remainder, subsequently adding 1 to its value. Lines 6 and 7 publish this number to the screen in two slightly different formats, on two separate lines.

Reading into the poem for its symbolic values, it is hard to tell if Warnell is moving well beyond the criticisms leveled against Winterbottom's work. Although the ability to remove Warnell's text and run it as code and therefore blur the line between the two languages is intriguing, Warnell doesn't stretch the use of the *Date()* object beyond convention. The *Date()* object is used to record the date (and time) but little else. Although Warnell seems to make a move forward, he falls within the same realm of criticism as Winterbottom.

Returning to Cayley, who argues that on-screen code is text, not code, the question at hand is the following: How do we study text and code as distinct species of thought and communication? How, in other words, do we develop a study of digital writing that recognizes interplay and upholds the distinction between the two species? From what disciplinary perspectives are code and text equatable? Whereas some scholars are more interested in the critical reception of a given project, others are more interested in how a given text is produced. I don't have the answer to this question, but in what follows, I develop one possible response to a question that will be sure to "slip" depending on the context in which it is observed. I present brief studies of two projects that illuminate the difference between user-operational and machine-operational approaches to digital writing and serve as powerful examples of the ways in which text and code represent an increasingly nonlinear coordination of distinct writing practices.

Project 1: "Crushed under Sodium Lights"

Developed by writer Dave Bevan and artist-programmer Brendan Dawes, "Crushed under Sodium Lights" was published in 2002 (Figure 2). Their project is a provocative example of one approach to using ActionScript to create a relatively simple plane of temporal and kinetic expression to great effect.

On screen, the project is divided among eight individual scenes. Each scene includes several lines (or fragments of lines) of text. The relations among the words and phrases in each scene change as a user moves the mouse pointer across the space of the project. As the mouse pointer is moved, fragments of text become visible and appear to rotate along

Figure 2. Dave Bevan and Brendan Dawes, "Crushed under Sodium Lights"

one or more elliptical orbits, and this mouse-driven movement changes the relations among the words. In the sixth scene, the following words are made visible with mouse-driven interaction:

```
EXIT
follow a trail of film
Conversation
A
glimpse of wisdom
may be
```

On the visual plane, we can study the ways in which the capitalized portions of the text structure new meanings along a schematic parallelism. We can also study the way in which the word "conversation," which is yellow, restructures the focus of the text in the scene; the other words and phrases are white. Moreover, we can study the ways in which particular words become increasingly bright as they appear to move to the front of the screen, shifting along the elliptical Z-axis of the spatial-kinetic plane.

Along the spatial plane, we can study the ways in which new syntactical meanings emerge as individual words and phrases shift their place along the elliptical Z-axis. Since this spatial plane is accessed via the mouse, we might say that this plane emerges in its interaction with the kinetic. As we look at the effectShape() function that Dawes developed, we can intuit some of the ways in which this spatiokinetic plane is constructed.

```
1    function effectShape (mcName, origX, origY) {
2       var mcObj = _root[mcName];
3       // get the distance
4       var theDist = _root.calcDistance(mcObj._name);
5       mcObj._alpha = theDist;
6       mcObj._y = origY+random((theDist)/10);
7       mcObj._x = origX+random((theDist)/10);
8       mcObj._x = origX + _root.calcDistanc(mcObj._name);
```

```
 9      mcObj._y = origY + _root.calcDistance(mcObj._name);
10      mcObj.s.setVolume(theDist);
11   }
```

A brief analysis of the ways in which origX and origY are used will illustrate the ways in which code and text are coordinated. In line 1, the values for origX and origY are passed *from* somewhere else in the machinic process *to* this function. In my estimation, these two "arguments" represent the *x* and *y* values for the mouse pointer. In other words, *origX* and *origY* represent the two-dimensional, spatial position of the mouse pointer on-screen. Ultimately, the *x* and *y* values will be transformed with the help of the random() method and the numerical value of *theDist*. The new value that is generated will be used to reposition one of the textual objects on-screen, like *EXIT* or *glimpse of wisdom*. With the help of *random()*, the eleven-line *effectShape()* function distributes a user-generated mouse movement along a plane of numerical potentiality.

The *random()* function or method is a part of the Flash programming environment. It is one of the predefined methods from which a writer-designer can draw. The method is used to generate a single value from a range of numerical potentiality. When a numeric value is "passed" or placed within the parentheses of this method, the application generates a "random" value based on the range implied by the number. For example, *random(5)* will generate one of the following five values: 0, 1, 2, 3, or 4. If we assume that *theDist*, which is defined somewhere else in the machinic process, has a value of 50, then we can hypothesize the following process: when *theDist* is divided by 10, the *random()* method will generate a number between 0 and 4. This "random" number is then added to the value of *origY*, which will be used to reposition along the y-axis a textual object such as *EXIT*. In line 6, Dawes is adding (or subtracting) a randomly selected number from a range of values to *origY*. In line 9, the new value of *origY* is passed to *mcObj_y*, which we can assume is one of the textual objects on the screen.

What I find particularly interesting about the way Dawes wrote this function is that it doesn't use trigonometric functions to construct the orbits around which the words appear to revolve, which suggests that a programmer-writer has more than one path to success. In a number of books and online tutorials in 2001–2002, mathematical functions and algorithms emerged as a novel way to explore new topographies and new modes of relating objects on-screen. In *Flash Math Creativity*, trigonometric functions are used to create fluid movement in three dimensions, and calculus functions are used to create fractalized images. In response to an e-mail in which I asked him how this effect was programmed, Dawes responded, "I . . . play[ed] around with numbers until I got the desired effect I want[ed]" (par. 1). In Dawes's project, the orbits emerged from his own experimentations. In other words, the orbits are themselves an effect of a field of generative potentiality that Dawes "bricolaged" together. If there's a lesson to be learned from Dawes's *effectShape()* function, it is that writing on both sides of the presumed line between text and code is a process of invention.

Project 2: "Stark, North of Gainsboro"

Developed by programmer David Deraedt, designer Qubo Gas, and poet Joan Houlihan, "Stark, North of Gainsboro" was published in 2003. Using a three-stanza poem by Joan

Houlihan as the basis for their "postalphabetic" explorations, Deraedt, Gas, and Houlihan's project is a compelling example of writing "after the voice." Once their project begins, the three stanzas disappear from the screen, leaving three words: *landscape*, *tree*, and *Cold*. As the work progresses, the letters from the first two words break free completely from their semantic bind. Letters (such as the *a* from *landscape*) move erratically across a two-dimensional, spatiokinetic plane of expression, the range for which is scripted aleatorically by a number of small machinic processes.

From the machinic-operational point of view, the variables and homegrown functions that script the aleatoric plane of spatiokinetic movement are dispersed across a number of textual and imagistic objects on screen. There are bits of code attached to individual letters as well as to various images on screen. Juxtaposed with each of the letters that make up the word *landscape*, for example, the following lines of code can be read to appreciate what is occurring at the machinic-operational level:

```
onClipEvent(enterFrame){
this.move_randomly(_root.move_val, _root.vit_val, 600, 0, 600, 0);
}
```

The *onClipEvent(enterFrame)* method is another predefined component in the Flash application environment. It is triggered each time Flash replays the frame in which this particular letter is placed. The "default" number of iterations in a Flash work is twelve per second. Deraedt changed the number to sixty, which means the *move_randomly()* function that he wrote is replayed approximately 3,600 times over the sixty-second run of the project.

The *move randomly()* function is used to generate the aleatoric range of movement for the letters on screen. In the following excerpt of code from his *move_randomly()* function, we can study the way he has scripted the spatiokinetic plane:

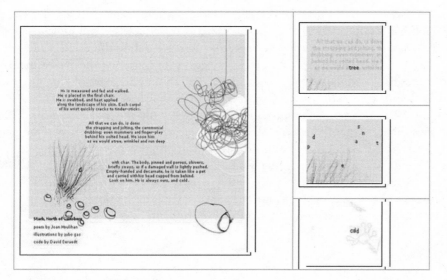

Figure 3.   Quobo Gas and David Deraedt, "Stark, North of Gainsboro"

```
1    move_randomly = function (move, vit, x_max, x_min, y_max, y_min) {
2        if(m==1){
3            if(this._x>x_max){
4                    this.val_x=-(vit);
5                    }
6        elseif(this._x<x_min){
7            this.val_x=vit;
8            }
9        else if(random(move)==1){
10            this.val_x=-this.val_x;
11            }
12        if(this._y>y_max){
13            this.val_y=-(vit);
14            }
15        else if(this._y<y_min){
16            this.val_y=vit;
17            }
18        else if(random(move)==1){
19            this.val_y=-this.val_y;
20            }
21        }

21    }
```

Thousands of times over the course of the project's play time, this function generates the movements for each of the individual letters on screen. There are a number of variables for which the values are assigned outside this function, so I will attempt to summarize what is happening with a few glosses. First, the function is passed six arguments or values. The *move* argument, which has a value of 20, is used in the random() method, which was described in the study of "Crushed under Sodium Lights." Then, the *vit* argument, which has a value of 1/10, defines the speed at which a letter will move. *Vit*, which is short for the French word *vitesse* (i.e., speed), and *move* define the velocity of each letter. The last four arguments (x_max, x_min, y_max, y_min) define the two-dimensional space of potential movement using x-y coordinates. In another part of the application, Deraedt defines the two-dimensional space as 600 (600 pixels.

Lines 3–12, which script the movement of a letter along the *x*-axis, and lines 12–20, which script the movement of a letter along the *y*-axis, are co-constitutive of a moment at which one of three events can occur: (1) the letter can progress positively along the x or y axis, (2) the letter can progress negatively along the x or y axis, or (3) the letter will stand still. If we substitute my use of the coordinating conjunction *or* with the "artificial" conditional phrasing *else-if*, we can appreciate the way a value tumbles down one or more conditions before satisfying a state that generates a movement.

Conclusion: Programming and (Digital) Poesis

Arguably, it is no accident that a chapter about generative writing highlights two e-poetry projects. In his book titled *Digital Poetics: The Making of E-Poetries*, Loss Peque_o Glazier argues that e-poetry may be "uniquely qualified to serve as the site for emergent forms of digital textuality" (95). Whereas hypertext provided humanities theorists and writing practitioners a simple "algorithm" (i.e., the link) with which to practice writing "after the voice," Glazier underscores the importance of poetry as a new agent of exploration and change: "[O]ne can argue that poetry, not prose, is the arena for this medium to be explored. . . . A true digital text is highly sensitive in every grain of its surface. This practically defines innnovative poetry, the ability for any word of character to be part of the action of the text" (95).

As programming languages continue to evolve and provide writers with increasingly "sensitive" dimensions of thought and expression, poetry may take over the vaunted position that prose occupied in first-generational explorations of hypertextual writing. When compared to the kinds of link-based textualities upon which early projects like Amerika's "GRAMMATRON" are based, projects like "Crushed under Sodium Lights" and "Stark, North of Gainsboro" can complicate the traditional notion of textuality. As online writerly thought and expression becomes increasingly sophisticated, prose might prove too bulky and unwieldly to navigate writing at the "zero-degree" of the pixelized grain. Programmers with an interest in digital writing may find that poetry provides a more subtle "application environment" in which to develop new modes of coordination between code and text.

In both "Crushed under Sodium Lights" and "Stark, North of Gainsboro," a "home-grown" aleatory function is used to help generate innovative interactions along postalphabetic dimensions of writing that include the spatial, visual, and kinetic. Compared to the hypertextual link, this interrelationality "after the voice" leads us to new levels of coordinated practices between text and code. In programming environments like Flash, the two species of writing will be coordinated in ever-changing ways, leading to new avenues of both cataloging and theorizing electronic textualities. As writer-programmers find new ways to program texts along every pixelized grain of its surface, the very notion of a text will evolve.

Notes

1.   Turkle talks about the competing aesthetics of computing represented by DOS and Mac enthusiasts. The DOS environment is largely transparent, and its users are enamored by the ability to access the coded textualities that inform the interface. Mac enthusiasts, on the other hand, are less concerned with depth than with the beauty of Mac's screen-deep, simulated appearance; Mac enthusiasts are not concerned with what Andrews calls the "*'neath textualities*" of the programmed environment.

2.   This focus on the performative or generative dimensions of code and their subsequent link to poetry echoes part of an argument developed by Rita Raley in her article "Reveal Codes: Hypertext and Performance."

3.   Even while scholars like N. Katherine Hayles call on compositionists to delve more "deeply into the machine," "high level" languages like ActionScript are a useful threshold: we can delve more "deeply" while retaining a connection to the human language arts on which our fields are based.

Works Cited

Andrews, Jim. "[empyre] Welcome Jim Andrews Re: Electronic Poetry." E-mail. 3 May 2003. *Empyre—Soft Skinned Space.* http://lists.cofa.unsw.edu.au/pipermail/empyre/2003-May/001336.html.

Cayley, John. "The Code is not Text (unless it is the Text)." *Electronic Book Review.* 10 Sept. 2002 http://www.electronicbookreview.com/thread/electropoetics/literal.

Cox, Geoff, Alex McLean, and Adrian Ward. "The Aesthetics of Generative Code." *Generative.net.* 1 July 2000 http://www.generative.net/papers/aesthetics.

Cramer, Florian. "Program Code Poetry." *Netzliteratur.net.* 1 July 2004 http://www.netzliteratur.net/cramer/programm.htm.

Dawes, Brendan, and Dave Bevan. "Crushed under Sodium Lights." *Born Magazine: Art and Literature Collaboration.* 1 July 2004 http://www.bornmagazine.org/projects/crushed.

Deraedt, David. Qubo Gas, and Joan Houlihan. "Stark, North of Gainsboro." *Born Magazine: Art and Literature Collaboration.* 1 July 2004 http://www.bornmagazine.org/projects/stark.

Glazier, Loss Peque_o. *Digital Poetics: The Making of E-Poetries.* Tuscaloosa: University of Alabama Press, 2002.

Hayles, N. Katherine. "Deeper into the Machine: Learning to Speak Digital." *Computers and Composition* 19.4 (December 2002): 371–86.

Houlihan, Joan. "Stark, North of Gainsboro." *Web Del Sol Association.* 1 July 2004 http://www.webdelsol.com/LITARTS/Joan_Houlihan/j-5.htm.

Knuth, Donald E. *Literate Programming.* Stanford, Calif.: Center for the Study of Language and Information, 1992.

Peters, Keith, Manny Tan, and Jamie MacDonald. *Flash Math Creativity: Second Edition.* Berkeley, CA: Friends of Ed, 2004.

Raley, Rita. "Reveal Codes: Hypertext and Performance." *Postmodern Culture* 12.1 (Sept. 2001) http://www.iath.virginia.edu/pmc/contents.all.html.

———. "Interferences: [Net.Writing] and the Practice of Codework." *Electronic Book Review.* 8 Sept. 2002 http://www.electronicbookreview.com/thread/electropoetics/net.writing.

Turkle, Sherry. *Life on the Screen: Identity in the Age of the Internet.* Boston: Simon and Schuster, 1997.

Warnell, Ted, and Nari Warnell. "LASCAUX.SYMBOL.IC." *Our Digital Lascaux: Hunters and Gatherers.* 1 July 2004 http://www.heelstone.com/lascaux/warnell.html.

Small Tech and Cultural Contexts

Overhearing: The Intimate Life of Cell Phones

Jenny Edbauer Rice

Can you hear me now? Good.

— Verizon Wireless advertisement

The scene is nearly always the same. A man in a jumpsuit walks around with his cell phone, checking reception hotspots in every corner of the earth. He asks his invisible caller, "Can you hear me *now*?" He pauses for the silent answer and then smiles. "Good." People around him catch his eye and hold up two fingers in an ambiguous peace sign cum victory sign cum *V* for Verizon. This scene makes the question seem strangely ambiguous: to whom exactly is he posing his question? It is not only the invisible interlocutor who hears the Verizon man. His bystanders also *hear him now*—they overhear him loud and clear. The slogan drips over into parody every day as I watch people with cell phones running from place to place while shouting, "Is *this* better? Can you hear me *now*?" Of course, the fact that we can all *hear them now* has produced a growing social irritation with cell phone usage in public. As Nedra Reynolds writes, "Listeners get annoyed when overhearing 'private' cell phone conversations because such conversations blur the lines between public and private space—concepts that most people would prefer to keep neatly separate even as technologies . . . make them increasingly blurry notions" (24). Hearing "private conversation" in "public places" thus disturbs a felt boundary between the two spheres.

But we would be wrong to presume that cell phones simply bring the private into public or make the public a private space. Cell phones exist as a liminal tool that collapses a distinction between public and private. Indeed, the technology actually works by creating what we might call a *zone of public intimacy*. It operates through the technical apparatus of *over*hearing. Cell phones are beginning to be used as a dating tool, for example, thanks to a service that notifies subscribers when a potential partner is within range. Users store their personal information and their dating preferences in their phones. When two "matches" are in close range, the two phones exchange information. As one industry leader explains this new dating trend, "When proximity computing meets social networking, very interesting things will happen"

(Biever par 5). Yet this kind of networking depends upon a sociality that is neither public nor private. Any merger of "proximity computing" and "social networking" is a technologization of another kind of hearing. It is neither a private nor a public hearing, but rather it is an intimate overhearing that occurs in technology's cellular space.

This is not only the case in the dating world but also in any number of social networks. Mass protests around the world are increasingly coordinated through the use of cell phone technology. As one activist posted to an online discussion of political activism and technology: "In the last decade there has been an upsurge in the political power of indigenous movements in the Andes which have their power base in rural mostly disconnected communities. A lot of that upsurge is due to the many years of organizing by indigenous leaders, social movements, and NGO's. That said, cell phones have acted as a major amplifier of their work. Increasing the ability for people to coordinate their actions and build robust social networks" (Rabble). Other protests have been successful specifically for their ability to utilize quick communications provided by cell phones. Plans can be coordinated on the spot, in real time, which builds in a high degree of flexibility.

Like the proximity networking of cell dating services, these grassroots protests are stretching the limits of public and private past any meaningful distinction. It is not merely that the private is in service of the public, nor is it the case that the private is becoming a function of the public. Rather, an intimate space of *betweening* is created by a technology "heard" both privately and publicly. Any distinction between the two—public and private—does not hold. As our Verizon man's question continually reminds us, cell phone conversations are rarely "heard" by the recipient alone. We *all* hear because we are held together by various kinds of proximities. The communicational lines of cell phones inevitably serve as a reminder of *the bleed* between boundaries that we still tend to think of as impermeable (as savvy as we may be about the death of the subject). As Brian Massumi explains, such public intimacy is sociality beyond what we normally propose as "the social." "There is the unbiddenness of qualitative overspill," he writes. "There is self-activity qualitatively expressed, presenting an affective order that is not yet 'yours' or 'mine'" (218). In short, the public and the private bleed together in those experiences of overspill.

Cell phone technology is, therefore, what we might call an apparatus of public intimacy. It opens up new spaces: new ways of being together and new ways of acting together. We are no longer hamstrung (if we *ever* were) between "the public spheres" or "the private spheres." Rather, new kinds of sociality mean "hearing" communications in ambiguous ways. "Like all good tools," writes Reynolds, "cell phones can be adapted to different situations and purposes. . . . [They] offer a means of resistance for users—even in such dramatic ways as calling from a highjacked plane to report the terrifying events that otherwise would have been contained within the jet" (23). In many ways, then, the intended recipient of our Verizon man's question—"Can you hear me now?"—does not matter. We all answer *yes*. I can hear you, hear your conversation to another, and hear myself being pulled into your private realm of communication. Though I am not a part of your private conversation, I *over*hear you. The sociality of your call *over*spills its social container and creates a space of public intimacy. Much like the Verizon man who is heard by both his interlocutor and bystanders, the cell phone user is neither fully public nor fully private. Rather, she exists as an intimate public body.

Works Cited

Biever, Celeste. "The dating game goes wireless." *New Scientist* 20 Mar. 2004 http://www.eurekalert .org/pub_releases/2004-03/ns-tdg031704.php.

Massumi, Brian. *Parables for the Virtual: Movement, Affect, Sensation.* Durham, N.C.: Duke University Press, 2002.

Rabble. "Cellphones, Rural Social Movements and the Bolivian Gas War." *Anarchogeek* 18 Nov. 2003 http://www.anarchogeek.com/archives/000256.html.

Reynolds, Nedra. *Geographies of Writing: Inhabiting Places and Encountering Difference.* Carbondale: Southern Illinois University Press, 2004.

I Am a DJ, I Am What I Say:
The Rise of Podcasting

Paul Cesarini

Why should we care about the iPod? It is just another MP3 player, another consumer electronics gadget designed to siphon wallets dry while users attempt to bask in its techno-utopian glow. The iPod wasn't the first such player, either, and it certainly won't be the last. It is expensive. It lacks features found in competing devices. It relies on a proprietary digital rights management (DRM) technology for commercial audio content, making it little more than an island in a sea of Windows-based products and services. There's really not much to the iPod, is there? The answer is both yes and no. While the iPod is typically billed as just an MP3 player, it provides significantly more functionality and has since been extended to even more uses, allowing people to express themselves and to disseminate their views in ways unimagined even a decade ago.

In addition to playing digital audio, the iPod also serves as an external hard drive capable of storing hundreds of audio and video files, presentations, papers, projects, assignments, and so on. iPod photos provide seamless audio-video integration with televisions and home entertainment systems. Add-on peripherals give the iPod digital audio recording capability, flash memory card reader support, and a host of other functions far removed from what we typically associate with MP3 players. The iPod is also bootable, at least for Mac OS X–based systems that allow for USB booting. If Apple finally chooses to release the near-mythical "Home on iPod" feature that was originally slated for release with Mac OS X 10.3, iPods could become even more useful. This feature would allow virtually any host computer to become your own, complete with your own bookmarks, preferences, desktop applications, and settings "no matter where you happen to be," according to a description that was briefly available on the Apple Web site. The idea was that users could simply "plug in the iPod, log in," and then be "home" in terms of their own personal computing environment. After doing so, according to this same fleeting information, users could return to their home computer and "synchronize any changes . . . by using File Sync, which automatically updates offline changes to the user's home directory" (McLean par. 3). Since iPods are roughly the size of a deck of playing cards and can easily fit inside a shirt pocket, using these or similar devices could redefine how students use institutional systems. The iPod's flexible, user-centric focus contributes to its high degree of popularity and visibility. In a recent interview with Leander Kahney, Michael Bull, a lecturer in media and cultural studies at the University of Sussex, stressed that the iPod "gives [iPod users] control of the journey, the timing of the journey and the space they are moving through. . . . The main use [of the iPod] is control. People like to be in control. They are controlling their space, their time and their interaction" (Kahney par. 17).

And then there is podcasting. The concept is comparatively simple: podcasting is essentially an audio version of an RSS ("really simple syndication" or "rich site summary") feed, designed to synch to an iPod or similar device each time the user synchs their device to their host computer, assuming they have opted to manually subscribe to the podcast in question ("What Is Podcasting?" par. 2).

When it comes to podcasting, there is no Federal Communications Commission (FCC). There is no Recording Industry Association of America (RIAA). There is no Clear Channel, Infinity Broadcasting, or Viacom. There is no AM or FM spectrum, since none is being used. There is no payola, no artificial push to promote Ashlee Simpson or the latest pop music flavor of month. In short, there are few if any intermediaries in the traditional sense. When it comes to podcasting, there are only podcasters, who produce and record their content and disseminate it by way of free or inexpensive software such as podcast cocreator and former MTV VJ Adam Curry's iPodder, Profcast, iTunes, Garageband, and numerous others. And there are only podcastees, who ultimately decide whether or not to subscribe to a given podcast feed. There are podcasts that likely generate a wide appeal and have legions of listeners, such as Curry's "Daily Source Code" or the News.com daily technology news. There are less well-known podcasts, such as David Tarleton's "Creative Mythology," which features original music and thoughts from, you guessed it, David Tarleton.

Certainly, the capability to record and upload a digital audio file, with the intention of having others access it, is nothing new. Yet prior to podcasting there was no free, open standard to push these audio files to computers and then portable devices. Before podcasting was developed, if you wanted to listen to a specific interview or audio program, you had to either turn on the radio or launch a Web browser and go to the respective site. It is now possible to make that content truly mobile, to listen to it whenever or wherever, Internet connection or not, and have it updated each time the device is synched. We can now listen to movie reviews while jogging, Rush Limbaugh while flying, or National Public Radio (NPR) while diving (assuming a watertight case accompanies the iPod in question). My students can listen to me, courtesy of my course-related podcasts, while they go to the gym, drive, or lounge around (so far, few actually do listen, but the number is growing each semester). They can even watch my video podcasts, which typically include presentation slides, images, audio narration, me as a "talking head," and related media.

These additional levels of actual or potential functionality inherent in iPods have not gone unnoticed institutionally. Duke University began a much-hyped pilot program issuing iPods to all incoming freshmen (Dean par. 1). The University of

Western Australia (UWA) did the same to all students in their communications studies program. Similar efforts are also underway at Georgia College and State University (Sellers, par. 2). In each case, students will be using their iPods to access course lectures and related content including audio eBooks and institution-specific calendar events and contacts. Students will also be using these as external hard drives to store and present various multimedia projects (Fardon, par. 5).

Of course, given the necessarily fluid nature inherent in most information technologies, it is difficult to predict how the iPod might survive as a viable classroom technology, much less how the iPod might survive increasing competition from rival consumer electronics manufacturers. These devices and evolving uses for them are simply too new to pin down with any degree of certainty how people will or will not use and interact with them. Despite no effective demand for such a service, the iPod has proved to be precisely the sort of liberating technology people craved. In 2003, there were no podcasts. In 2005, there were thousands of them, with new ones being added each day. Apple CEO Steve Jobs has even integrated podcasts directly into iTunes (Louderback, par. 2).

Yet this is the very late age of print that Jay David Bolter, Raymond Kurzweil, and others have forecasted for so long, blended with an even tighter media convergence. What might have previously been a text-based Web page or blog now becomes an on-demand audio blog, with or without still images and video, not broadcasted as we have come to expect from analog and Internet radio, but instead available via a synched feed to an iPod or similar device to as many or as few people who care to listen. Given the potentially revolutionary role podcasting could play in the coming years by allowing basically anyone to become their own radio station— free to express their own personal or political views, free to express their own musical tastes—and given the broad appeal the iPod and subsequent iPod culture continues to have on our students and in education, I suppose a better question would be, why shouldn't we care about the iPod?

Works Cited

Bolter, Jay David. *Writing Space: The Computer, Hypertext, and the History of Writing*. Hillsdale, N.J.: Lawrence Erlbaum, 1991.

Dean, Katie. "Duke Gives iPods to Freshmen." *Wired News* 20 July 2004 http://www.wired.com/news/digiwood/0,1412,64282,00.html.

Fardon, Mike. "Ipods Come to Class at UWA." *University of Western Australia* 10 Mar. 2003 http://www.uwa.edu.au/media/statements/2003/march/ipods_come_to_class_at_uwa_(10_march).

Kahney, Leander. "Bull Session with Professor IPod." *Wired News* 25 Feb. 2004 http://www.wired.com/news/mac/1,62396-1.html.

Louderback, Jim. "Apple's Jobs Announce iTunes Podcast Support." *eWeek* 23 May 2005 http://www.eweek.com/article2/0,1895,1818895,00.asp.

McLean, Prince. "Apple's missing 'Home on iPod' features resurfaces in filing." *Apple Insider*. 11 Oct 2006 http://www.appleinsider.com/articles/06/10/11/apples_missing_home_on_ipod_feature_resurfaces_in_filing.html.

Sellers, Dennis. "University Finds Innovative uses for iPods." *MacCentral* 28 Apr. 2003 http://maccentral.macworld.com/news/2003/04/28/ipodinnovation.

"What Is Podcasting?" *Podcast Alley* 31 May 2005 http://www.podcastalley.com/what_is_a_podcast.php.

Walking with Texts: Using PDAs to Manage Textual Information

Jason Swarts

All manner of texts are built from visual, verbal, and multimedia representations of organizational and individual knowledge. These texts help form an organization's public face, and they are tools for managing work activity. Texts and the technologies that support their creation and distribution have developed as artifacts of systematic management (Yates). Obvious examples are texts like standard operating procedures, but texts like inspection reports and patient histories also contain conventional representations of information that support increasingly standardized work practices. Supporting technologies like typewriters and photocopiers make these texts easier to produce and distribute within an organization. The personal digital assistant (PDA) falls on this same developmental trajectory. However, the device's design as a personal and mobile piece of information technology has facilitated its use in a different kind of text-mediated activity, one that works around standardization while supporting it.

Increasingly, people recognize the need to receive some textual information on their PDAs to support work practices that take them away from their desks and homes (Churchill and Munro). Often the information needed for such work is embedded in existing texts, written in generic conventions that assume standard work practices. PDAs mediate work activity at a very local level, where activity is far messier and more ad hoc than standard operations and processes can describe or anticipate (Brown and Duguid 95–97). The textual information required in these situations must be flexible enough to be repurposed to fit the demands of the situation, which often cannot be foreknown (see Suchman's notion of "situated actions"). Texts and technological interfaces that constrain adaptive reuse of texts hamper local work practices out of which the work of an organization emerges (Henderson 465–66).

The conflict is that texts are technologies of place—they help create order and stability by which places are defined. But places are comprised of spaces, which are temporary "intersection[s] of mobile elements," activities that collectively give rise to a coherent place "actuated by the ensemble of movements deployed within it" (de

Certeau 117). PDAs are spatial technologies that mediate these day-to-day activities that collectively comprise an apparent stability and order. Using Michel de Certeau's ponderous example, cities are comprised of the intersecting paths and motives of their inhabitants, whose lives act out the city. The daily activities of walking, buying, eating, and greeting (i.e., spatial practices) all compose the city and its public spaces, restaurants, shops, and squares by means of what de Certeau labels a "walking rhetoric" (101). Likewise, piecemeal uses of text and adaptive application of that information to local work situations collectively actuates the work of an organization. PDAs will ideally support such spatial practices but only if we envision them as more than text delivery devices.

Consider a doctor and nurse who share patient information with the aim of developing a patient's "treatment plan," a composite text whose form is built from an ad hoc arrangement of information drawn from a variety of texts. The development of the treatment plan requires improvisational use of patient information from a variety of texts, some of which are designed to support unrelated work practices (e.g., billing records) but contain information that can be repurposed. The staff's ability to make the treatment plan is tied to their ability to interact with and not just receive texts out of which the treatment plan is constructed. This use of patient texts illustrates de Certeau's two principal walking rhetorical moves: asyndeton, in which parts of texts are separated, and synecdoche, in which parts are assembled again as a totality with situated relevance (101). Both rhetorical practices are typical of everyday uses of text (Duchastel 179).

Texts are metonymic representations of organizational places. They are arrangements of information that enact and support the activities by which organizational work is defined. Their use imposes order and standardization from above (Yates 64). But this order is just as often an emergent quality of many separate spatial practices. The value of PDAs as spatial technologies is in their ability to remediate organizational texts as locally valuable pieces of textual information through which users can stroll and assemble new texts to meet immediate and perhaps temporary needs.

Developers of PDA technology have started to recognize that PDAs can be effective tools for managing the messiness of text-mediated work and creating conditions that allow organizations to emerge from local practices. One PDA application that offers useful mediation is "Total Recall," a system that couples a PDA with an electronic whiteboard to capture information recorded on the whiteboard. Information written on the whiteboard is stored as XY coordinates on a desktop computer. When needed, the coordinates are sent via wireless connection to a PDA, where the erased information is reconstructed as an image. As a person waves a PDA over the surface of the whiteboard, the PDA screen displays past versions of content in their original locations (Holmquist). Enhancements have included a jog dial that allows users to browse whiteboard content at different points in time (Sanneblad), allowing users to walk through the whiteboard text both spatially and temporally.

Users of "Total Recall" have access to a variety of textual information that they can stitch together and reconfigure to suit the demands of the situation. Because "Total Recall" allows users to revisit transitional ideas and give them visual form, users can construct a temporary text built out of texts that are distributed in time. The effect is that "Total Recall" facilitates a new form of composition that crosses time boundaries and mimics, more closely, nonlinear uses of text in many cooperative work settings.

The current popularity of using PDAs to mediate local text-mediated work practices suggests an immediate need for us to think critically about the mediating role that PDAs will play in managing the volume of textual information present in those activities. Analyzing text use as a practice of walking with information and through information-scapes promises to be a powerful heuristic.

Works Cited

Brown, John Seeley, and Paul Duguid. *The Social Life of Information*. Cambridge, Mass.: Harvard University Press, 2000.

Churchill, Elizabeth F., and Alan Munro. "WORK/PLACE: Mobile Technologies and Arenas of Activity." *SIGGROUP Bulletin* 22.3 (2001): 3–9.

de Certeau, Michel. *The Practice of Everyday Life*. Berkeley: University of California Press, 1984.

Duchastel, Philip C. "Textual Display Techniques." *The Technology of Text*. Ed. D. H. Jonassen. Englewood Cliffs, N.J.: Educational Technology, 1982. 167–92.

Henderson, Kathryn. "Flexible Sketches and Inflexible Databases: Visual Communication, Conscription Devices, and Boundary Objects in Design Engineering." *Science, Technology, and Human Values* 16.4 (1991): 448–73.

Holmquist, Lars Erik, Johan Sanneblad, and Lalya Gaye. "Total Recall: In-place Viewing of Captured Whiteboard Annotations." *Extended Abstracts of CHI 2003*. 2003. http://www.viktoria.se/fal/ publications/2003/totalrecall.pdf.

Sanneblad, Johan, and Lars Erik Holmquist. "Browsing Captured Whiteboard Sessions Using a Handheld Display and Jog Dial." *Future Applications Lab—Viktoria Institute*. 2003 http://www .viktoria.se/fal/publications.html.

Suchman, Lucy. *Plans and Situated Actions*. New York: Cambridge University Press, 1987.

Yates, JoAnn. *Control through Communication: The Rise of System in American Management*. Baltimore: John Hopkins University Press, 1993.

Text Messaging: Rhetoric in a New Keypad

Wendy Warren Austin

Our creativity surges when opportunities for maintaining human contact present themselves. Thus, when cell phone companies began offering text messaging services to customers, intending mainly to help businesses, cell phone users co-opted the feature for themselves. Today, more than one in four Americans use text messaging (Lenhart and Shiu), an amazing example, as Huatong Sun points out, of the public's unexpected adaptation of a technological device for purposes other than those originally intended.

Why is text messaging becoming so popular? It's a silent, convenient, and affordable way to communicate with others. It's not a very simple way to write, however. Nor do most cell phones have an indefinite amount of space for what can be written: 160 characters, to be exact, are permitted per message on most American-made cell phones. If I simply want to respond to a text message with a yes, I must press 99922777 (which gives me the three characters *Y*, *E*, and *S*) and the button marked "send." But perhaps it is the ease with which I can reply to the silent message that makes text messaging appealing. One Sunday night as I sat on my couch grading papers, my cell phone on the end table beeped once. When I picked it up, I saw that Justin from my first-year writing class had sent this message: "Can I get an extension for my paper til Wed.?" I tapped back, "sure, I extended it to everyone anyway til Wed," and set my small phone back on the table, returning to my papers. I commented to my spouse on how amazing it was to be able to communicate with my students whenever they needed, wherever I was or they were. It feels like when I used to pass secret notes to my school friends. Alex Taylor and Richard Harper note that young people see text messaging that way, too, as a way of exchanging gifts. Special. Unexpected. Paper airplane notes conveyed by radio towers.

Apparently, text messaging was an afterthought feature discovered by the phone companies and marketed as an added bonus. Only 160 characters are possible to send at one time because that is the typical space left over at the end of the bandwidth used for other purposes (Hård af Segerstad 187). As text messaging was used more, instant messaging software made it possible for instant messages sent over the Internet to be forwarded to cell phones as text messages. If you are not at your computer, you can still stay in touch with others through your mobile phone. Constant contact—do we want that? Apparently, one fourth of Americans, one third of Chinese, half of Finns, and millions of Swedes, Germans, Brits, and Indians do, and the list goes on. Besides, you can always turn your phone off if you feel like being alone.

Computer keyboards were somewhat familiar to all of us as we began to use computers because we had typewriters from which to migrate. But a keypad? Now, that's a new way to think of writing. My keyboard has eighty-seven keys on it, including the twenty-six letters of the English alphabet. My cell phone has twenty-three buttons on it, only twelve of which allow me to "write." It's like writing in code: A = 2, B = 22, C = 222, D = 3, and so on. (Maybe this appeals to me so much because I wanted to be a secret agent when I was a kid and used to hide encoded messages all around the house for my brothers.) To write an "s," you have to tap the seven key four times. It would be interesting to find out whether those who take quickly to text messaging also have a propensity for spy shows or cryptograms. Though some people take to texting easily, some do not. Once a person learns how to enter text, it doesn't seem like such a bother to type in those coded messages. Some can even type messages under a table without even looking. Again, the secret-note mystique attaches itself to this activity.

Even with the unfamiliar writing interface and coded symbols, more people every day are embracing cell phone text messaging. James E. Katz and Mark Aakus propose the term *apparatgeist*, which loosely translates to "machine and mind or spirit," to describe how human behavior is transformed by machines (11). Medical doctors (Bauer et al.; Sherry, Colloridi, and Warnke) and therapists (Hilaire) are using text messaging to keep in contact with their patients. Russians are using it to drum up Communist support (Gutterman).[1] Teachers are using it to stay in touch with students (as in my own example) and their teacher trainees (Seppälä and Alamäki). Parents are using it to stay in touch with their children. Teenagers and young adults, as always happens with new technology, are using it more than most, sometimes so much that it affects their sleep habits (Van den Bulck). They text others to keep in touch with friends, to exclude them (Smith and Williams), to bully them (Keith and Martin), to flirt with each other, to find friends in a crowd, to convey emotions, to cheat on tests (Powell), and to do virtually everything with a communication device that they are able to do with 160 characters at a time.[2]

The limitations of the cell phone keypad and the coding necessity lend itself to an alternate means of expanding the possibilities of use. Edna Aphek notes that besides using basic alphabetic literacy, younger generations are also becoming adept at photo-visual literacy, the ability to see the symbols we use for meaning in their shapes as well as in the symbols they represent. Take smileys, for instance. If I type a colon, a hyphen, and a back slash, :-\ means that I am dismayed or confused; 8=D represents a person with glasses smiling widely. Necessity is the mother of invention, as the saying goes. Therefore, in the interest of brevity, the rest of this article will be in text message-speak.

We r nterin a new stj in wrtng. Uv cors th rjistr in wich we spEk is dfrnt here: ths is n akadmek rticl whil th sel fOn is n infrml venu. But they ovrlap, as u kn C here. W8, tho, wil they ovrlap 2 mch? Hrd 2 tl. MayB thts wht tEchrs r 4.[3]

Notes

1. Text messaging is referred to in the cell phone industry as SMS (short messaging service). Text messages are known in Russia as "esemeski" (Gutterman).
2. See Crispin Thurlow's work for a thorough analysis of a range of language functions with SMS.
3. We are entering a new stage in writing. Of course, the register in which we speak is different here: this is an academic article while the cell phone is an informal venue. But they overlap, as you can see here. Wait, though, will they overlap too much? Hard to tell. Maybe that's what teachers are for.

Works Cited

Aphek, Edna. "Digital, 'Highly Connected' Children: Implications for Education." Plenary talk at The Seeing, Understanding, Learning in the Mobile Age Conference. Budapest, Hungary. April 2005. 13 May 2005 http://www.fil.hu/mobil/2005/Aphek.pdf.

Bauer, S., R. Percevic, E. Okon, R. Meermann, and H. Kordy. "Use of Text Messaging in the Aftercare of Patients with Bulimia Nervosa." *European Eating Disorders Review* 11 (2003): 279–90.

Gutterman, Steve. "Russian Communists Turn to Text Messaging." *High Technology CustomWire* 15 Mar. 2005.

Hård af Segerstad, Ylva. "Use and Adaptation of Written Language to the Conditions of Computer-Mediated Communication." Diss. Göteberg University, Sweden, 2002.

Hilaire, Yvonne. "Texting Clients Who Are Victims of Domestic Violence." *Counseling and Psychotherapy Journal* 15.4 (May 2004): 7.

Katz, James E., and Mark Aakhus, eds. "Introduction." *Perpetual Contact: Mobile Communication, Private Talk, Public Performance.* Cambridge: Cambridge University Press, 2002.

Keith, Susan, and Michelle E. Martin. "Cyber-Bullying: Creating a Culture of Respect in a Cyber World." *Reclaiming Children and Youth* 13.4 (Winter 2005): 224–28.

Lenhart, Amanda, and Eulynn Shiu. "How Americans Use Instant Messaging." *Pew Internet and American Life Project Report.* 1 Sept. 2004 http://207.21.232.103/pdfs/PIP_Instantmessage _Report.pdf.

Powell, William. "An A for Effort." *TD* May 2003: 28.

Seppälä, P., and H. Alamäki. "Mobile Learning in Teacher Training." *Journal of Computer Assisted Learning* 19 (2003): 330–35.

Sherry, Eugene, Beatrice Colloridi, and Patrick H. Warnke. "Short Message Service (SMS): A Useful Communication Tool for Surgeons." *ANZ Journal of Surgery* 72 (2002): 369.

Smith, Anita, and Kipling D. Williams. "R U There? Ostracism by Cell Phone Text Messages." *Group Dynamics: Theory, Research and Practice* 8.4 (2004): 291–301.

Sun, Huatong. "Expanding the Scope of Localization: A Cultural Usability Perspective on Mobile Text Messaging Use in American and Chinese Contexts." Diss. Rensselaer Polytechnic Institute, 2004.

Taylor, Alex S., and Richard Harper. "The Gift of the *Gab*?: A Design-Oriented Sociology of Young People's Use of Mobiles." *Computer-Supported Cooperative Work* 12.3 (2003): 267–96.

Thurlow, Crispin, with Alex Brown. "Generation Txt? The Socio-linguistics of Young People's Text-Messaging." *Discourse Analysis Online* 2003. 19 April 2005 http://www.shu.ac.uk/daol/articles/ v1/n1/a3/thurlow2002003-01.html.

Van den Bulck, Jan. "Text Messaging as a Cause of Sleep Interruption in Adolescents, Evidence from a Cross-Sectional Study." *Journal of Sleep Research* 12 (2003): 263.

Beyond Napster: Peer-to-Peer Technology and Network Culture

Michael Pennell

> In a nutshell, peer-to-peer technology is god-
> damn wicked. It's esoteric. It's unstoppable. It's
> way, way cool.
>
> —Cory Doctorow, "The Gnomes of San Jose"

Although most widely recognized in the form of Napster or other music and media based file-exchange programs, peer-to-peer (P2P) technology is much more than a way for college students to avoid paying for music. It is distributed computing, allowing users to interface with not just a single computer or server but a constantly changing network, or ecology, of computers, servers, and small technologies. As Cory Doctorow claims, P2P may simply be "unstoppable," despite significant attempts to stop it by organizations such as the Recording Industry Association of America (RIAA) and academic institutions. At the very least, P2P networking is definitely "way, way cool." While still in its infancy, P2P technology will continue simultaneously to affect and reflect a significant societal shift—a shift toward what many term "network society" or "network culture" organized around movement (see Castells, *The Rise of the Network Society*; Taylor).

Looking Under Napster's Hood

Prior to Napster, SETI@home may have been the most popular and extensive example of P2P technology at work. Utilizing the downtime of machines connected to the Internet, SETI@home involves users downloading free software that will analyze radio telescope data when the computer is not in use in the search for extraterrestial activity. As David Barkai explains, "The idea behind P2P computing [such as Seti@home] is that each peer, i.e., each participating computer, can act both as a client and as a server in the context of some application" (4). More recently, this model has manifested itself in Internet-based applications in which users exchange files directly. In the case of Napster, or similar applications such as Gnutella or Kazaa, users download a networking program and maintain a collection of files—music, in this case—on their machines. Users may then search and download music from other users' collections, resulting in a decentralized network of peers. On college campuses throughout the country, populated by financially strapped music lovers with high-speed Internet connections, P2P networking, especially in the form of Napster, exploded during the years 2000–2001. According to Howard Rheingold, "At its height, 70 million [Napster] users were trading 2.7 billion files per month" (72).

P2P and Network Culture

However, as Barkai states, "Peer-to-peer is much more than sharing music files" (xii). P2P networking is about changing social arrangements. Napster and SETI@home are, or were, not the height of P2P possibilities. As Rheingold contends, "The main problem with Napster, from a P2P purist's perspective, is that it was not designed to be truly a decentralized network" (73). Users had to go through a central server to find music on others' disks. Future P2P efforts aim for an arrangement where contributing files or bandwidth are not optional but rather built into the network. Rheingold, building upon Doctorow, envisions the future of P2P networks as "a commons where the sheep shit grass, where every user provisions the resource he consumes" (80).

In this sense, P2P networking provides a unique melding of small technology and social networks. As Rheingold comments on the Napster epidemic, "The social network multiplied the impact of the network of computer storage" (72). P2P both requires and facilitates the networking of social relationships, leading to a completely decentralized structure. More than simply software or hardware, P2P represents a changing "mindset" (Barkai xvi). As Rheingold concedes, "People don't just participate in [P2P]—they *believe* in it" (65). In turn, technology companies, especially Apple, as well as numerous Web 2.0 applications, are attempting to work the P2P mindset of decentralized networks into their technologies (Kahney). Tim O'Reilly, praising Apple's move in this direction, sees the company "tak[ing] that network concept, the idea of reaching beyond the single device, and they're starting to build that into applications" (qtd. in Kahney). With the increase in wireless networks and portable devices, users do not interface with a single machine in a single setting. Movement is built into the interface.

In outlining the expanding network society, Manuel Castells posits, "the power of flows takes precedence over the flows of power" (*The Rise of the Network Society* 469). P2P, and the P2P mindset, reflects the power of flow—and lawsuits prove ineffective and misguided as they operate on an outdated idea of flow of power depending on fixed payment extraction points (Jones). Bonnie Nardi and Vicki O'Day, in their depiction of network ecologies, point to this shift toward the power of flow and movement: "The part we often focus our attention on is the technology: computers, networking, applications, handheld information gadgets, instruments, monitors, widgets ad infinitum. . . . But it is in *the spaces between these things*—where people move from place to place, talk, carry pieces of paper, type, play messages, pick up the telephone, send faxes, have meetings, and go to lunch—that critical and often invisible things happen" (66). Lawsuits and surveillance are visible reactions to seemingly invisible changes and flows. But it is in the in-between spaces, in the liminal zones between places and activities, where things are happening, where new social arrangements are developing.

These emerging social configurations are not tied to specific places or pieces of technology—there is no center. Each roaming person with earbuds connected to an MP3 player is constantly recreating ambient space; every user or participant is both a center and a node in the network. As with Napster, iPods, or Google, the technology was not the only key to the interface's success; rather, it was driven by a social arrangement. In describing Web 2.0, Tim O'Reilly maintains that, for example, "Google happens in the space between browser and search engine and destination content server, as an enabler or middleman between the user and his or her online experience." There is a *belief* functioning as the glue in the network—distributed computing just reflects the shifting social arrangement. Hierarchy is erased or, at the least, flattened in the network. But again, the network is only as strong as its participants' belief—representing a profound change from a hierarchical structure of control held together by fear (see the recent controversy surrounding Wikipedia versus Encyclopedia Britannica, as an example). While it may be admittedly utopian to see P2P networking as the key to a more democratic future, this is a movement with implications for the future of democracy and an engaged citizenry. After all, this technology evolves, in the words of Castells, not only "away from the halls of power . . . in [the] back alleys of society" (*The Power of Identity* 362) but also in the computer labs and dorm rooms of today's college campuses. But, more importantly, it is happening—it is developing—in the spaces between these places.

Works Cited

Barkai, David. *Peer-to-Peer Computing: Technologies for Sharing and Collaborating on the Net*. Hillsboro, Ore.: Intel Press, 2002.

Castells, Manuel. *The Power of Identity*. London: Blackwell, 1997.

———. *The Rise of the Network Society*. London: Blackwell, 1996.

Doctorow, Cory. "My Date with the Gnomes of San Jose." *Mindjack Magazine* 15 Oct. 2000. 3 Nov. 2003 http://www.minjack.com/feature/p2p.html.

Jones, Steve. "Music that Moves: Popular Music, Distribution and Network Technologies." *Cultural Studies* 16.2 (2002): 213–32.

Kahney, Leander. "Looking Toward a Networked World." *Wired* 29 Oct. 2003 http://www.wired.com/news/mac/0,2125,60999,00.html.

Nardi, Bonnie A., and Vicki L. O'Day. *Information Ecologies: Using Technology with Heart*. Cambridge, Mass.: MIT Press, 1999.

O'Reilly, Tim. "What is Web 2.0? Design Patterns and Business Models for the Next Generation of Software." Sept. 30, 2005 http://www.oreillynet.com/lpt/a/6228.

Rheingold, Howard. *Smart Mobs: The Next Social Revolution*. Cambridge, Mass.: Perseus, 2002.

Taylor, Mark C. *The Moment of Complexity: Emerging Network Culture*. Chicago: University of Chicago Press, 2003.

Communication Breakdown:
The Postmodern Space of Google

Johndan Johnson-Eilola

Cause I've got a golden ticket
I've got a golden chance to make my way

And with a golden ticket, it's a golden day.

—"I've Got a Golden Ticket,"
Willy Wonka and the Chocolate Factory

The information age promised us a lot of things; indeed, it seemed to promise us everything. Ironically, the commodification of information means that information, in itself, is worth very little. We should not be surprised. Snake oil and patent medicines always held out illusory promises. In the original film adaptation Roald Dahl's novel *Willy Wonka and the Chocolate Factory*, Charlie's Golden Ticket (Figure 1)

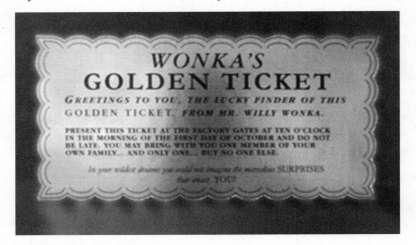

Figure 1. Willy Wonka's golden ticket

promises the moon in a golden day but only provides an opening, an opportunity, "a golden chance." The Golden Ticket seemed like an answer, but it wasn't: it was merely an opening, a way in. Getting out was much more complicated, as Veruca Salt discovered. We are changed, as an initial Google search reveals (see Figure 2).

Most of our search results are signifiers for a different Veruca Salt, the alt-rock band named (postironically) after the character in Dahl's novel—context once (or more) removed. Let's try again and hope there's not a band named "Willie Wonka" (see Figure 3).

Great! We're in business. Not so quick; the cached images are promising, but image is apparently as fleeting as text (see Figure 4).

Well, it's around somewhere. The domain has been bought and is now up for sale. Although Google tells us helpfully that "below is the image in its original context" (see Figure 5).

It seems pretty clear that that "original context" is not, as Derrida would say, "recoverable."

Our New Information Space

This appears to be an information management problem, although it's not. What we are seeing are the emergent conditions of a new information space, a shifting and contingent *datacloud*. We can categorize and fix the structure and content of such a space, but doing so inhibits our own motions within it and freezes the information

Figure 2.   Google image search on Veruca Salt (ca. 2003)

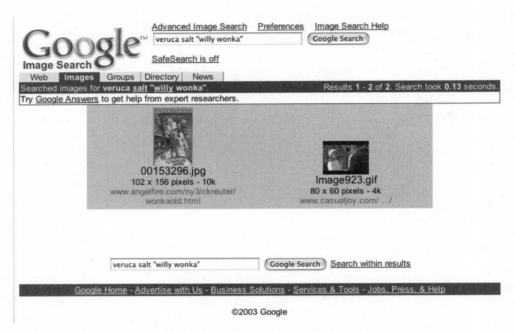

Figure 3.   Google image search on Veruca Salt "Willy Wonka" (ca. 2003)

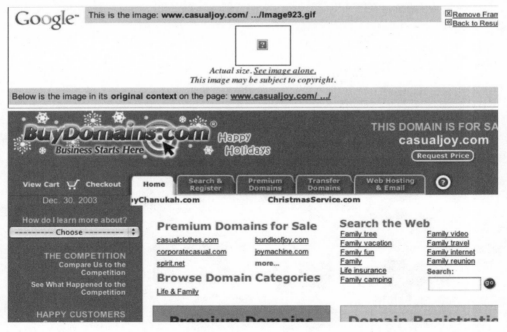

Figure 4. Orphaned results page for the search in Figure 3

> Below is the image in its **original context** on the page: **www.casualjoy.com/ .../**

Figure 5.   Loss of original context

and literally prevents communication; that is, paradoxically, when we communicate we're attempting to take some meaning out of our own idiosyncratic and specific experiences and transfer that meaning to someone else's idiosyncratic and specific experiences. But obviously it's impossible to do that perfectly: perfect communication would require that the sender's and receiver's idiosyncratic and specific experiences were absolutely identical. In the end, we can't even communicate perfectly with ourselves. Heraclitus wasn't sure exactly what he meant even as his words echoed into silence: You can't step into the same river twice; as Cratylus added: You can't even step in the same river once.

What River? What Words? What Context?

But that's OK. In fact, it's preferable. Communication is fundamentally about productive chaos: experiences jostle against each other, infecting us and mutating. So context both allows and inhibits signification. For example, what I mean by *Web*, based as it is at least partially on my own unique experiences, cannot be the same as what you mean by *Web* even if they may share some similarities. So we gladly decontextualize the texts that we share, or fragment them from their original contexts in order to create our own shared meanings: texts hook to other texts in an endless chain of signification.

This may seem like a bad thing, but it's not. It's how communication works.

> There is no difference what a book talks about and how it is made. Therefore, a book has no object. As an assemblage, a book has only itself, in connection with other assemblages and in relation to other bodies without organs. We will never ask what a book means, as signified or signifier; we will not look for anything to understand in it. We will ask what it functions with, in connection with what other things it does or does not transmit intensities, in which other multiplicities its own are inserted and metamorphosed, and with what bodies without organs it can converge. A book exists only through the outside and on the outside. A book itself is a little machine. (Deleuze and Guattari 4)

Contrary to what you might think, this loss of context is an important, valuable thing, for communication relies on continual partial loss of context. Meanings cannot haul their contexts around with them; the semantic baggage involved would crush them, make them immobile. Instead, they shed contexts and connotations in order to move. They travel light, but that does not make communication powerless.

Hierarchy is similar. At least some of the preference for Google over hierarchical indices such as Yahoo! lies in the information-age promise made by Google: *It's a wide-open field*. Compare the structural assumptions made by the pages in Figures 6 and 7.

Where Yahoo! constructs a logical, navigable information space, Google offers a flat surface. Where Yahoo! provides order, Google suggests chaos spread thin. In other words, where Yahoo! communicates in modernist terms, Google connects in postmodernist ways.[1] There are, similarly, two ways to read Google: from a modernist perspective and from a postmodernist perspective. Information architects and teachers tend toward the modernist perspective: "But the oracle—recently described

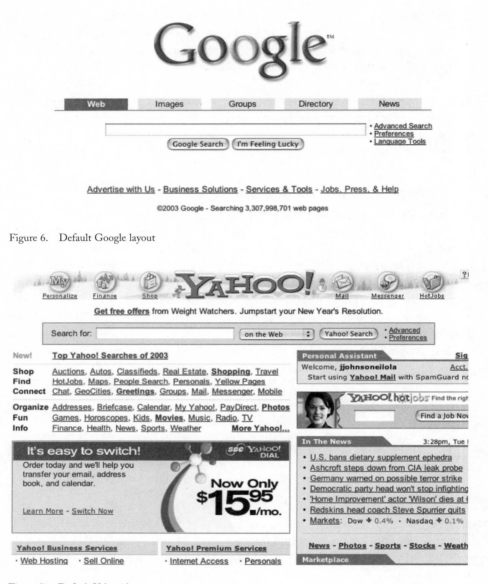

Figure 6.   Default Google layout

Figure 7.   Default Yahoo! layout

as 'a little bit like God' in the *New York Times*—is not perfect. Certain types of requests foil the Google search system or produce results that frustrate more than satisfy. These are systemic problems, not isolated ones; you can reproduce them again and again. The algorithms that Google's search engine relies on have been brilliantly optimized for most types of information requests, but sometimes that optimization backfires. That's when you find yourself in a Googlehole" (Johnson n.p.).

And Stephen Johnson is absolutely correct: Google doesn't contain everything And the bits that rise to the top in searches are often the *wrong* bits—from a modernist perspective, that is.

From a postmodernist perspective, communication breakdown is an important—a crucial—aspect of communication itself. Of *course* Google doesn't contain everything; of *course* Google doesn't return exactly what I want in the top ten hits of a search. Those gaps are where communication occurs; they are part and parcel of the experience.

So what Google excels at is this flattening of information space, the decontextualizing of information in order to make it more fluid. Although there's always the option to view search results "in context"—indeed, clicking URLs in the search results page is the normal step users take immediately following a search—that context is illusory and must be understood as always contingent. Google's foray into mobile technologies such as Google Local complicate this contingency further, decontextualizing and recontextualizing physical location to provide cell phone users with maps and directions on their tiny displays. As with traditional Google searches, results are often sketchy, more suggestive than exact. And as Google services mutate and combine with other custom applications through Google's open API, Google's contexts multiply even as they break down.

By breaking the control of the hierarchy and coming in through the back door, Google performs the sacrilegious act of making hierarchy secondary or incidental. Instead, we rely on the multitude of connections and the necessary loss of context in which we can create new meanings. According to Brandy Alexander, the transsexual hero(ine) of Chuck Palahniuk's *Invisible Monsters*, "Our real discoveries come from chaos," Brandy yells, "from going to the place that looks wrong and stupid and foolish" (258).

Note

1.   In fact, the success of Google has apparently spurred Yahoo! to revise their search pages to make the hierarchical search a secondary option. The default search from their main page gives users results of a Google-like text search rather than a hierarchical search. Not to be outdone, Google's much-lauded elegant interface has now mutated into various other possibilities: Advanced Search, Google News, Google Maps, and Froogle.

Works Cited

Deleuze, Gilles, and Félix Guattari. "Rhizome." *A Thousand Plateaus: Capitalism and Schizophrenia*. Trans. Brian Massumi. Minneapolis: University of Minnesota Press, 1987. 3–25.
Johnson, Stephen. "Digging for Googleholes." *Slate* July 16, 2003. 21 May 2004 http://www.slate .com/id/2085668/.
Palahniuk, Chuck. *Invisible Monsters*. New York: W. W. Norton, 1999.
Willie Wonka and the Chocolate Factory. Dir. Mel Stuart. Warner Studios, 1971.

Let There Be Light in the Digital Darkroom: Digital Ecologies and the New Photography

Robert A. Emmons Jr.

The postphotographic era, initiated by the invention of the digital photographic process, establishes new social, cultural, and ecological practices affecting the viewer, the artifact, and the creator, ultimately resulting in a new period of artistic expression. The postphotographic era is a result of advancement. Humans exist in an environment of "instants." "Now" refers not to an approximate figure defining a general period of time but to an exact moment, second, or instant. Technological progress is a means to an instant: instant coffee, instant information, instant download, instant playback, instant photography. "Instant" technology changes human movement and interactivity. "Instant" culture has aided in the posthuman environment, whose ultimate purpose is to enhance and improve human ability. This change, however, also results in a loss. For the photographer, the loss of spontaneity, serendipity, or accident creates a scenario wherein the eye is preceded by the computer in the chain of interpretation and creation. The ratio between the human eye and the camera lens establishes a new creative dynamic. Therefore, photographers see the digital camera not as an extension of the eye but as an extension of the computer both in the production and post-production of images. The postphotographic world, then, is a wholly new photographic ecology. Its components can no longer function as before. Consequently, digital photographers belong to this new era because they cannot make the same choices or photographs they once made with a traditional camera.

In trying to understand the context of the postphotographic image-object, photographers must investigate the new relationships among their bodies, the subject, the environment, the image, and the digital tools in both a production and post-production environment. This line of investigation can begin by examining the primary digital tools in digital photography: a digital camera (production) and desktop digital imaging software such as Adobe Photoshop (post-production).

The primary difference between the traditional photographic camera and the digital camera is the ability to instantly screen the photographed image. Before "the screen" became an integrated part of the digital camera, it was the photographer's mind that preplayed then played back the image. When discussing method and technique, Minor White consistently and eloquently expresses the relationship between the photographer's camera, eye, and mind in relationship to his subject: "To engage a sequence we keep in mind the photographs on either side of the one in our eye" (*Mirrors, Messages, Manifestations* 123). White also describes the photographer's mind as a piece of film, a blank sheet. But this is a special kind of blankness. It is an active state of mind, a receptive state of mind, ready at an instant to grasp an image but with no image preformed in it at any time. The lack of a preformed pattern or preconceived idea of how anything ought to look is essential to this blank condition. Like a sheet of film, the mind is seemingly inert yet so sensitive that a fraction of a second's exposure conceives a life in it (White, *Camera Mind* 17).

The photographer must previsualize what the final image will look like. He or she uses experience and tools to interpret what the light and chemicals would produce on the film in the camera. This is not to suggest that photographic knowledge and technique are unnecessary or unneeded in digital photography, but it certainly suggests the implications of the photographer's relationship with his or her photographic subject and environment. When a photographer foregoes previsualization for instant playback, there are a number of changes in the way images are made. For the photographer, there is the constant recognition of the immediacy of the image and a heightened reflexivity of the process.

When a photographer uses a traditional 35-millimeter camera to photograph a subject, something happens. A relationship is formed. There is the photographer and the photographed. This exchange works like a dialogue between artist and subject and is dictated by every passing instant. With the camera to his or her eye, spontaneity begins to force itself and a stream of conscious photography ensues. The photographer is trying to capture the essence of the subject through various forms of communication, distances, and functions. Again, White establishes the intimacy and urgency of photography's spontaneous relationship between artist and subject: "Creativity with [photographs] involves the invocation of a state of rapport when only a camera stands between two people . . . mutual vulnerability and mutual trust" (*Mirrors, Messages, Manifestations* 195). With a digital camera, these conditions exist but tend to develop in a much different way.

In the postphotographic era, the digital camera can no longer be Henri Cartier-Bresson's "extension of [the] eye" (1). When the photographer removes the camera from the eye, the extension is broken; the monitor becomes the interface of experiencing the picture, not the camera. The observed is re-observed, and the subject becomes removed from the eye and is now observed through the computer aboard the camera. The postphotographic environment establishes a postocular (i.e., beyond the eye) theory of photography. Digital photography presents new photographic seeing that ultimately affects how the photographer chooses to make the final object. The stakes at the moment of capture are no longer as high as they once were. With the knowledge of what is correctable, editable, and in general fixable, the photographer can walk away from a subject sooner and with more confidence that he or she will get the final print through post-production.

Traditional post-production exists in the lab or darkroom, which has long held a mystical quality. Out of the darkness and liquid, from precious metal to paper, comes something magical. The traditional darkroom is a kinetic process that involves and needs the organic qualities that exist between artist and art object. When developing and printing, a photographer must handle the negative, the paper, the chemicals, and the tools. The body must handle the process: working with enlargers, taking the paper out of the box without creasing it, putting it into the easel and squaring the blades, rocking the trays, and transferring the paper to different baths. It is a process of movement that establishes a very physical relationship. It is a method that is idiosyncratic and involves particular and peculiar tricks of the trade to produce the desired piece. Creating traditional prints is an experience of simultaneously dancing, challenging, manipulating, and ultimately painting with light.

Unlike the traditional darkroom, the digital darkroom is not dependent on the environment. The traditional darkroom needs sinks, drains, vents, tables, and trays to shut out and manipulate light. In the digital darkroom, there is no manipulation of light. Instead the photographer needs computers, scanners, software, and printers to manipulate numbers, bits, and pixels. The relationship between the photographer and his or her piece is not physically possible until it is pulled from the printer. It is a virtual association—behind a wall or a screen—and must be negotiated as a third party. The photographer is moving the mouse, selecting the tools, and telling Photoshop to dodge, burn, crop, and mask, but he or she is one step removed from the actual object and tools. This slight disassociation is the fundamental challenge photographers must overcome to connect with their work as they traditionally did. Thus the kinetic process becomes more static. Movement is confined to the boundaries of the mouse pad. Photoshop becomes the darkroom, and it creates a paradigm shift in how a product is conceived and produced, involving all affecting factors from economical to artistic.

Photoshop imitates the process and nomenclature of the traditional darkroom but provides speed, precision, and editorial components not available in the traditional darkroom. Working digitally provides the artist an unerring hand. Mistakes can be undone with the set "edit/undo" strokes, and multiple versions and layers can be handled simultaneously for comparisons and generations of a photograph. Simultaneity is the very essence of the computer as a "dynamic screen" (Manovich 96). This ability leaves little margin for error. The unpredictably and serendipity that is occasionally experienced in a traditional, final print emerging from the tray is largely nonexistent. Chemical photography exists because of the nature of elements and their more fluid interactions. Spontaneity is sparse in the postphotographic era because it arises from a synthetic space and the more exacting actions that the space makes possible. Again, as with the digital camera, the same work cannot be created in the digital darkroom as the traditional darkroom. From its actual makeup to its path of creation, the final object will always be different.

Works Cited

Cartier-Bresson, Henri. *The Decisive Moment*. New York: Simon and Schuster, 1952.
Manovich, Lev. *The Language of New Media*. Cambridge, Mass.: MIT Press, 2001.
White, Minor. "The Camera Mind and Eye." *Magazine of Art* 45 (1946): 16–19.
———. *Mirrors, Messages, Manifestations*. New York: Aperture, 1969.

"A Demonstration of Practice":
The Real Presence of Digital Video

Veronique Chance

Presence is only presence at a distance and this distance is absolute, that is, irreducible.

—Maurice Blanchot, *Friendship*

[T]here is no true presence in the world—in one's own world sense of experience—other than through the intermediary of the egocentron of a living present; in other words, through the existence of one's own body living in the here and now.

—Paul Virilio, *Open Sky*

As digital technologies are made more accessible and available to the artist and as the proliferation of (media) images through the use of such technologies increasingly define our culture, art practices are being redefined, and our notions of and relationship to the visual are being reexamined. Through the use of digital technologies such as the Internet, digital video cameras, and data projectors, our relationship to visuality and to vision itself is being questioned. These visual technologies are allowing artists to refine and redefine our relationships to the visual, to challenge and alter our perceptions of reality, and to examine our relationships with ideas and modes of representation in which questions concerning issues of presence and nonpresence, proximity and distance, past and present, and public and private become increasingly blurred.

"A Demonstration of Practice" was an event I set up and performed at Goldsmiths College, London, on February 21, 2005. The performance was meant to describe such a process and methodology of my art practice, which examines these relationships through the use of digital technologies, particularly those (technologies) of the still and moving image. The demonstration combined elements of live performance, performance to the camera, and the representation of performance to highlight the relationship between the screen and physical (bodily) presence (see Figure 1). Crucial to the event was the presence and role of the digital video camera, a small but potent device with a big role to play, which acted as catalyst to the whole situation. A centerpiece through which all elements connected and combined, it was there to witness, record, and represent a live act while simultaneously being a part of the live act itself.

Strategically placed in direct correspondence with both the body and the act taking place, and positioned in relation to other elements that formed part of the whole construct, the camera acted as a prop, as both a receiving and transmitting device, and as a sculptural object. Not only would it perform an act in the presence of an audience (in its live recording and representation on screen), but also, while on display, it would be admired for its aesthetic qualities as much as for its technical capabilities: for its smallness and portability, for its metallic cleanliness and shininess, and for its fetishistic status as an art object. As part of the act, it was there to be observed and to seduce the audience as much as the image it would transmit with the aid of a data projector. Paul Virilio writes of an audience's "passion for gazing" and of "a vision industry based entirely on the motor" (89). Here these would come into relation as the audience's gaze would rest not only on the performance and its representation as an electronic image but also on the producer and transmitter of that image: on the camera itself.

The performance itself took the form of a ten-minute workout sequence performed on a silver exercise ball. Bringing this event to a public arena put me, the performer, on public display. Being watched and recorded in the present meant that the

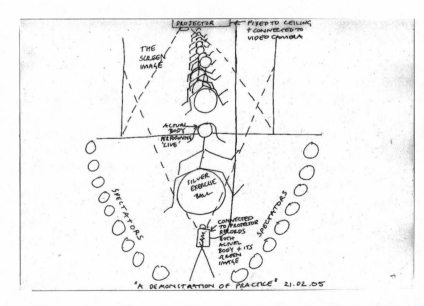

Figure 1. Diagram of layout: "The position of the camera in relation to the performing body, its image, and the audience." Copyright 2005, Veronique Chance.

physicality of the body was emphasized. Not only was the audience made aware of the physical presence of the body as it exercised in front of them, but as a performer, I was very much aware of their presence and of my own presence as I was performing. By putting myself on display, I was aware that my every move would be watched by an audience of spectators as well as by the camera. I became intensely aware of the following actions and perceptions: the movements I was making, of the rhythm of those movements, of the transition from one movement to another, of not making a mistake, of where I was looking, of my breathing, of my heart rate, of my physical contact with the ball and with the floor, and of the fact that these movements in themselves, through their rhythm and in their repetition, simulated sexual activity.

The physicality of the body was also emphasized by its representation as a projected image on the wall directly behind the performance arena. Proximity and distance were both reinforced and contradicted in a situation where the physical closeness of the live body created a psychological distance from the spectator, who did not like to get too close. The screen image, physically at a distance, was there to seduce its viewers, simultaneously inviting them closer whilst also keeping them away. The "passion for gazing" thus suggested not only an act of looking and seduction but also an act of looking and seducing at a distance.

This situation was further reinforced by the fact that, by simultaneously recording the live act and projecting the images in real time, the camera had picked up both what is in front of it and what has just been recorded. In a seeming temporal convergence and a temporal impossibility of recording both past and present at the same time, the screen image became one of an infinite repetition of the act it was recording. As each repeated image receded further and further back to infinity, a

Figure 2. Still from performance: "The physicality of the body was emphasized by its representation as a projected image directly onto the wall behind it . . . the screen image became an infinite repetition of the act it was recording." (Copyright 2005, Veronique Chance).

temporal delay of movement also occurred as the present was continuously picked up by the camera and thrown back onto itself again and again, creating not only a physical but a temporal distance. This became all the more seductive, however, as the rhythm of the repeated movements themselves were further emphasized in the continuous representation of the moving body, itself in repetition.

With the presence of the camera as part of the act, the audience was able to directly compare the physical act itself with its representation as an image simultaneously and to see that they were different, that the screen image as a representation—with all the breathing, the sweat, the dirt from the floor, and the squeaking of the ball as the body came into contact with it—did not represent the reality of the physical body performing in front of them. The projected image cleaned up all of this. This is why they may have been more comfortable watching the image itself, despite its physical distance, rather than the physicality of the real body performing right in front of them, which was perhaps too real and too close and too intimate in a space that was too open.

The proximity of the physical body to the audience made the relationship between them one of enforced intimacy, and it was this enforced intimacy that actually ended up creating a distance again. There was a kind of invisible barrier between the audience as spectators and the performing body, because they were watching a private act of which they were not a part. Although the nature of the live act gave them permission to watch and legitimized them as spectators, they were physically removed from it, and what was intimate on one level actually became less intimate on another. The act in itself was one of a kind of intimacy with itself in the sense that it was the exposure of what would have normally remained within the confines and privacy of an artist's studio presented in a very direct way. This sense of privacy somehow remained so that the spectators were not sure how to encounter it and it became something more confrontational. Even though the physical act confronted the spectators, the camera presented them with a representation of that act. In the form of an image, the act was more palatable: the physical distance between the image and its audience made it something they did not have to confront—they could still be seduced by it while at the same time keeping their distance.

In an age that is becoming more and more media-oriented, our perceptions and encounters with the physical body are increasingly experienced through cameras and screens. Through an encounter between the physical body and its screen image, "A Demonstration of Practice" suggested that it may be more difficult to look at and to contemplate the physicality of a real body, even when it is presented right in front of us, and that a physically immediate situation is something to which audiences are becoming less accustomed.

As a performer, I am able to experience, feel, and encounter the presence of my own body as a *living present* as I am performing. But as spectators, perhaps we can now only experience or encounter the presence of the body *at a distance*, that is, at a distance that has now become *irreducible*. Perhaps our "passion for gazing" and our passion for technologies of vision have together created a passion for technological distance at the expense of a passion for physical (bodily) proximity. And perhaps this encounter at a distance is now, paradoxically, a more authentic experience for us. As an artist, what the presence of the camera allows me to experience, feel, and explore however, is the presence of both of these types of encounters at the same time and to re-present them as present acts of presence in the present.

Since this chapter was written, "A Demonstration of Practice" has been performed in various forms at symposia and conferences, both nationally and internationally.

Works Cited

Blanchot, Maurice. *Friendship*. Trans. Elizabeth Rottenberg. Stanford, Calif.: Stanford University Press,1997.
Virilio, Paul. *Open Sky*. Trans. Julie Rose. New York: Verso, 1997.

Buffering Bergson: Matter and Memory in 3D Games

Julian Oliver

> I find, first of all, that I pass from state to state.
>
> —Henri Bergson, *Creative Evolution*

Henri Bergson would have liked 3D computer games, or at least the way they are made. Naturally, this may seem unusual, since the field of 3D graphics is itself the grandchild of Euclid's *Elements*, a geometric construction of the universe as a mesh of points connected by measurable lines (Mueller). However, contemporary 3D games deploy a range of techniques in their presentation that differ greatly from the alienable universe as it is mathematically observed. At the center of the 3D game there lives a conspicuous, albeit unlikely echo of Bergson's (sometimes less than popular) ideas.

Bergson believed in a kind of creative evolution whereby the perceived world is itself in a state of perpetual change, a morphology produced at the intersection of memory and actions derived from experience (*Matter and Memory*). Bergson countered the idea that this state of change is outside the subject; rather, he said, the subject produces change through action itself. This annoyed Bergson's contemporary, Bertrand Russell. In a famous showdown, Russell insisted that there is no such thing as a "state of change" so much as a scientifically determinable series of observable states.

Russell was an empiricist, so he was frustrated by Bergson's notion that time exists as a continuum of perceived durations, a self-preserving past that inevitably converges upon a mutable present: "In reality, the past is preserved by itself automatically. In its entirety, probably, it follows us at every instant; all that we have felt, thought and willed from our earliest infancy is there, leaning over the present which is about to join it, pressing against the portals of consciousness that would fain leave it outside" (Bergson, *Creative Evolution* 5). From this perspective, Russell's universe is comparable to the frame-by-frame transformations in a 3D game, each a state of transformation relative to the last. But in Russell's world, the user is innately independent from the unfolding of the world itself. It is here that Bergson's morphological universe finds company in the architecture of a 3D game engine.

Right from the origins of the video game as a flat field of interactive "primitives"—basic geometric forms such as planes, cubes, spheres, cones, and cylinders—to the rich mediascapes of the contemporary game, one artifact remains intact: the in-game camera (Figure 1). Both a trigonometric construction and the hemisphere around which action is grouped, this window to the game world delimits the periphery of our immediate capacity while also serving as the context from which we chart our vectors in play. At first it seems the world of games somehow preexists the camera—that regardless of whether or not we're playing, the shape of the world itself persists platonically. However, this is not the case. The very mathematical construction of this camera and how the world is drawn around it defines a peculiar and even metaphysical difference between the worlds of game and life. In a 3D game, it is the viewer who produces the world. When I say "produces," I don't mean "makes" or even puts in a shiny box so much as "ushers" or "brings into effect."

In a 3D game, the camera deploys what is called a "projection matrix," which can be thought of as a geometric keel against which all visible objects are shaped and sheared. Each object in a 3D game is comprised of many differently oriented triangles that comprise "meshes." These meshes can be skewed during play to give the illusion of depth that is drawn relative to the orientation of the camera. There is, in fact, no "depth" in a 3D world, just a preferential stacking of triangles drawn around the attention of the software camera. This order of drawing is done in what's called the "Z buffer," where Z is the axis of representative depth. When we are looking at

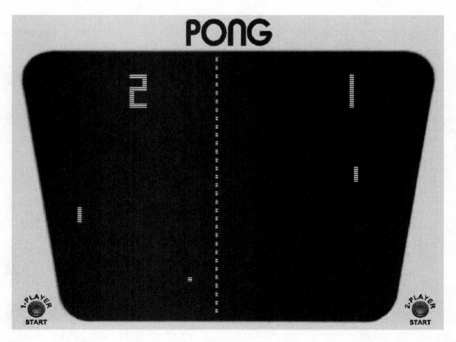

Figure 1.   Still from the game *Pong*. Basic forms ("primitives") comprise the elements drawn to the screen.

a scene of hills, giant killer goats, and an inviting bridge in a 3D game, for instance, we are, in fact, looking at a flat surface of triangles, each angularly attenuated to give the appearance of form (Figure 2). Form itself, however, in the sense of separable, material integrity does not and has never existed in a 3D game. As the player moves the camera around the world, these triangles flex and foreshorten, some drawn on top of others and some with different surface effects, all of which help inversely affirm the player as the occupant of a visually distinct location.

But there is another question, not just of what is drawn and when, but of how much is drawn. "How much world can I see?" Here, 3D game development takes a very different approach to the construction of worlds—not out of conscious design but from an architectural necessity whose roots reach down to the system hardware itself.

Game data come in a variety of types. Some are persistent (e.g., user Interface, soundtracks, player character mesh, and items), while others are nonpersistent, which means they are encountered only when they are relevant to the current stage of game play (e.g., nonplayer characters, other players, a hill, a flying whale). As the camera is pushed around the world by either the human player at the input device or entities in the game, nonpersistent game data is "culled" from view and new data are drawn, current with the camera position. Parts unseen by the camera, like the back of objects or objects immediately behind the camera, are removed from view, giving preference to logically perceptible objects in a process called "backface culling" (Figure 3).

Another technique used (often concurrently) is "far clipping." This is best understood as a maximum viewable distance or hard limit of ocular perception

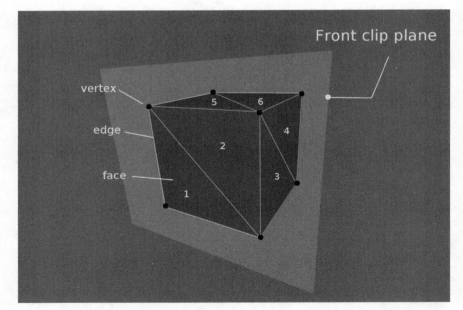

Figure 2.   The translation of six 3D triangles (faces) into the screen area of the client computer system, giving the illusion of a 3D cube.

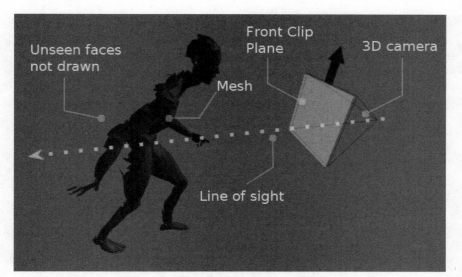

Figure 3.   The technique of "backface culling." Mesh "faces" (triangles) not seen from the perspective of the 3D camera are not drawn.

designed to draw only what is considered close enough to be relevant to the scene. The "far clip plane," together with the front or "near clip plane," comprise the "view frustrum," or the renderable region in a 3D world. These two technologies—"culling" and "clipping"—are designed precisely to restrict what can exist in the physical graphics memory of the gaming platform. Here the world, as a quantifiable universe of measured and positioned parts, is both held and shaped by the limits of memory.

In a 3D computer game, the subject both reveals and shapes the world as a surface of transformational effects from which we produce our own frames of recounted experience. We do not "create" the world in a 3D game as much as we evolve, manifest, and "disappear" its parts in and out of shared memory as a function of action itself. This configuration, wherein the perceiving subject is intrinsically bound to the unfolding of the world itself, is ulterior to the popular understanding of how our own corporeal world works—a persistent, indifferent enclosure that subjects merely visit.

It is therefore notable that the art of creating 3D worlds has itself derived from the traditions of mathematics and geometry, two sciences that seek to describe and project universal relationships in spite of the subject. Where strict computational simulation of the real is concerned, such languages of universal description have retained integrity (envisage a simulation of a car impacting with a wall or of sunlight acting on a glass). However, once a player is introduced with choices, actions, and the capacity to navigate, the technology must adapt to support a world that is produced through and for the subject.

Bergson was often asked, "so who does the remembering, us or the world?" In a game, and in Bergson's universe, it's always a bit of both.

Works Cited

Bergson, Henri. *Creative Evolution*. New York: Dover Publications, 1998.
———. *Matter and Memory*. New York: Zone Books, 1991.
Mueller, Ian. *Philosophy of Mathematics and Deductive Structure in Euclid's Elements*. Cambridge, Mass.: MIT Press, 1981.

Shifting Subjects in Locative Media

Teri Rueb

All the world's a stage.

—Shakespeare, *As You Like It*

A new area of narrative experimentation has emerged with the proliferation of mobile communications technology. The convergence of personal digital assistants (PDAs), pocket PCs, mobile phones, and portable music and video players with cellular, Wi-Fi, global positioning systems (GPS), and Bluetooth capabilities allows media content to be delivered at specific geographic locations, or in response to movement or relative proximity. Narrative works designed for delivery via such "locative media" seek to tell stories that unfold in real space. These works draw on dramatic, literary, cinematic, and spatial design conventions that have shaped the structure of narrative experience across cultures since ancient times. Narrative works in locative media are unique in that they engage the body of the participant as the agent that drives the narrative forward, revealing location-specific information through physical movement. The physical journey becomes an important structuring element in these narratives as movement is tightly coupled to the unfolding of events in time and space. In conjuring media through movement and framing it visually, spatially, and acoustically, the participant becomes a roving point-of-view that sits somewhere between the discursive space of the narrative and the physical space of the environment.

Some more common types of narrative-based locative media include automated interactive museum guides (iPod-based user-generated content), interactive historical reenactment dramas (*Riot*), or interactive real-space games (*Uncle Roy All Around You*). These examples draw on site-specific conditions to heighten interpretive, dramatic, or agonistic effects. Such works tend to engage the location as a foundation, skeleton, or concrete reference upon which the narratives sit as augmentation to the environment. "Place" is understood as a noun in such works, a given static condition upon which media content is hung as a kind of overlay.

While dynamic correspondences may occur between media, place, and meaning in such works, their narratives tend to have an internal logic independent of the place, moment, or movement of the participant through the piece. The participant's movements may serve to "drive" the work forward, but this participation is rarely considered integral to the emergence of meaning in the work. For example, even though each player may experience *Uncle Roy* differently, the general concept is that of a treasure hunt where the individual's movement and interaction with a particular place is not the central thread that brings meaning to the experience. Similarly, narrative cinema and literary forms, from plays and novels to film and hyperfiction, offer points of view that are predetermined either by the camera image itself as an indexical or symbolic reference to point-of-view, the proscenium stage, or the narrator's voice in text or voiceover. This is true in each instance regardless of the potential number of points of view contained by the script, database, or playback system or the number of points of view available simultaneously on screen, on stage, or in the soundtrack.

By contrast, a handful of narrative works in locative media have emerged that begin to explore storytelling forms and conventions that specifically exploit the highly indeterminate interaction of place, time, narrative, and the mobile body of the participant. Such works begin to form complex ecologies of signification in which meaning literally and metaphorically shifts with the movements of each individual participant. These works approach "place" as a verb, "movement" as a highly indeterminate choreography, and "point of view" as radically multiplied, fragmented, fluid, and unstable. Among these works are projects by Janet Cardiff, which employ linear soundtracks that subtly direct participants to locations with voiceover or visual directives, and by Stefan Schemat whose oeuvre has consisted solely of GPS-based narratives since the early 1990s. I have also experimented with GPS-based spatialized narratives since 1996, most recently in "Drift" (2004) and "Itinerant" (2005).[1]

Although examples of spatialized narratives that resist the conventions of the audio tour are few, they point to an emerging set of narrative conventions that are specific to locative media. Strong examples of narrative-based locative media tend to combine literary, cinematic, theatrical, and spatial design conventions to exploit the role of the participant as an essential player in the work. In such works, meaning only becomes fully manifest through the participants' physical movements as they weave together and fuse with the physical, virtual, and discursive elements of the composition.

Works that begin to articulate a formal language and syntax that is unique to locative media tend to exploit the specific site or place in such a way that the narrative and its structure and meaning emerge from and are dependent on the interaction of participant, place, time, and social context. Like hyperfiction and other nonlinear narrative forms, narrative works in locative media embrace a potentially indeterminate order of events, the incorporation of the participant's agency in driving the narrative forward through choice and the process of selection, and the potential for multiple narratives to unfold from a given set of constrained variables. However, locative media narratives go further in engaging audience members as essential players in establishing point-of-view, setting, time, character, and plot

Figure 1.   Listening to "Itinerant" at the threshold of Boston Common and the Downtown/Financial District of Boston

development—not just through making choices about route or path through the given material, but through their position as subjects who exist in physical, as well as discursive, contexts comprised of events, settings, and characters that are accounted for by the author as essential variables in shaping the work's meaning. The physically situated condition of the mobile participant is not taken for granted in these works, but is instead exploited as a means of establishing a complex ecology of variable characters, settings, times, and points of view.

The distinction between narrative-based locative media and more established storytelling genres is especially clear in the case of point of view. Locative media narratives are unique in their combination of multiple, indeterminate, fixed, and non-fixed points of view that simultaneously exist in virtual, discursive, and real space.

Unfolding in real space and in response to the body of a wandering participant *in situ*, narrative-based locative media offer multiple and simultaneous points of view

Figure 2.   Visitor exploring the shore of Wadden Sea, the site of "Water" by Stefan Schemat and "Drift" by Teri Rueb (Courtesy of Cuxhavener Kunstverein.)

that blend and are shaped by all the complex indeterminacy of the participant's agency and place within an actual and discursive milieux of social, cultural, physical, and natural forces. The body of the participant becomes an implicit mobile point-of-view that becomes fragmented by and fused with the myriad points-of-view offered through the narrative itself and the physical and cultural environment in which it unfolds. In this respect, narrative locative media works explicitly engage the participant as an embodied subject.

Despite historical links to the most ancient practices of oral tradition from walkabout to the peripatetic school of Aristotle, we are at a very early stage in the evolution of narrative-based locative media. Audiences need to be exposed to and educated in the thoughtful analysis and interpretation of a medium that combines narrative with movement, theater, large-scale spatial design, and everyday spaces. Eventually a language of narrative in locative media will emerge that is as rich and complex as the storytelling conventions we take for granted in theater, literature, and cinema. Close readings come from extended exposure and careful consideration of the historical, social, political, and aesthetic context in which a work exists. With time a sensitive audience and a body of critical reading practices will emerge that are specific to narrative work in locative media. Even as the industry and information have led us to see ourselves as ever more fragmented subjects with literary and cinematic traditions that express this condition, so will our increasingly mobile culture reveal new subjectivities and a complex set of storytelling conventions will evolve, perhaps establishing locative media narrative as a mature genre.

Note

1. Documentation of "Drift" is available at http://www.terirueb.net/drift. Documentation of "Itinerant" is available at http://turbulence.org/Works/itinerant/index.htm.

Future Technologies and Ambient Environments

9

Virtual Reality as a Teaching Tool: Learning by Configuring

James J. Sosnoski

If you have read St. Clair Drake and Horace H. Cayton's *Black Metropolis*, perhaps the most famous account of an African American urban neighborhood ever written, you know what happened over the years to Bronzeville in Chicago. Shortly after the turn of the twentieth century, African Americans migrated from the South in large numbers to seek employment in northern cities. Many came to Chicago attracted by the accounts of life there in the *Defender*, a widely circulated African American newspaper. In the '20s and '30s, Bronzeville was an entertainment district as famous and popular as Harlem in New York. But like its eastern counterpart, Bronzeville declined economically and, by the mid-century, became a ghetto housing the poor. Attempts to provide low-income housing in the area resulted in notorious "projects" where crime and drugs flourished amid run down homes and boarded storefronts. Then, in the last decade of the twentieth century, Chicago's economy shifted dramatically and residences replaced warehouses and factories or high-rise projects in areas surrounding the loop, including Bronzeville.

The irony is that real estate developers began to tear down sites that had immense significance in African American cultural heritage—the Regal Theater, for example. The once-famous jazz and blues theater that had hosted nearly every musician featured in the histories of jazz and blues no longer exists. All that is left are some blueprints, tapes, and photographs of famous musicians who played there, like Miles Davis, Louis Armstrong, and Fats Waller. Bronzeville's history is disappearing as the buildings in which it was centered are torn down and replaced with history-less monuments to urban rehab.

There is a group in Bronzeville that wants to teach the children growing up there about their vanishing African American heritage—about the *Defender*, the South Side Art Center, the Rosenwald, the Southerland Lounge, the Pilgrim Baptist church, and the architects, entrepreneurs, and politicians that made Bronzeville Second City Harlem. I was contacted by this group because of the work I had been doing on Virtual Harlem. They requested a showing in the University of Illinois at Chicago's (UIC) CAVE (computer-assisted virtual environment; rooms where visitor are immersed in a virtual environment pictured on its walls), a million dollar immersive virtual reality (VR) installation. After the showing, they wanted to pursue the possibility of creating a Virtual Bronzeville.

The Bronzeville group could not afford the type of immersive VR experience in which they first encountered the African American history associated with the New Negro Movement as it was formulated in Harlem, New York, in the '20s and '30s. Particularly attractive to the group was a version of Virtual Harlem that had been produced with a gaming engine and could be shown on a computer. Two members of the Arts and Sciences Collaborative Exchange Network Development project (ASCEND)—Bryan Carter in Columbia, Missouri, and Ken McAllister in Tucson, Arizona—were using gaming software and techniques in the development of teaching tools. In Carter's case, he wanted to make Virtual Harlem available to the widest possible audience. In McAllister's case, he wanted to organize a group of programmers to develop learning games.

Using gaming engines to exhibit historical cityscapes excited the Bronzeville group because Virtual Harlem and Virtual Bronzeville could be experienced in their neighborhood at a modest cost. They also believed that younger visitors to the proposed cultural center and information bureau would be far more likely to explore Virtual Bronzeville if it had the look and feel of a contemporary computer game. We also discussed long range plans to allow interested parties to download sections of the cityscape—say the block on which they lived—onto personal computers and eventually personal digital assistants (PDAs). Looking even further into the future, the ASCEND group anticipated Webcasting a live tour of Harlem using a wearable computer. The tour guide would point out the buildings that remained from the thirties as the viewer walked past sites such as the Cotton Club. Viewers would be able to toggle back and forth between virtual and actual Harlem and experience the unfolding of the neighborhood's history.

Since the Bronzeville group included educators from Northeastern Illinois University's Center for Inner City Studies, the design of the ASCEND Collaborative Learning Networks (CLNs) appealed to them because the CLNs are highly interactive and involve computer modeling of the history being studied (see Figure 1). The Virtual Harlem project is the prototype for ASCEND CLN structuring projects such as Virtual Bronzeville and Virtual Montmartre.

The Virtual Harlem Project

Since 1998, I have been working with a group of collaborators on the Virtual Harlem project, a VR learning environment shared by several universities and used in conjunction with courses on African American culture. Virtual Harlem is the brainchild of Bryan Carter, who initiated the project while a graduate student at the University of Missouri. As a teacher of courses in the Harlem Renaissance, he submitted a successful proposal to the university's Advanced Technology Center to build a VR model representing Harlem, New York, in the 1930s.

In 1998, when I first experienced Virtual Harlem, the project was far enough developed that I was able to walk down streets, listen to Langston Hughes recite a poem, overhear Marcus Garvey's speech, overhear two men playing checkers in an alleyway, enter the Cotton Club, watch Cab Calloway's performance, and take a trolley ride. Since then, I have been collaborating with Carter on various pedagogical experiments with Virtual Harlem.

During the spring and winter semesters of 1999, Virtual Harlem was the centerpiece of courses in the Harlem Renaissance taught in parallel at UIC and Central Missouri

Figure 1.   The start of an immersive tour of Virtual Harlem

State University (CMSU). Students followed similar syllabi, communicated with Course-Info, a part of the Blackboard software, and at critical junctures visited Virtual Harlem. After each visit, we arranged for a video conference session in which both classes could see and talk to each other, though the teachers did most of the talking (Sosnoski and Carter, *Virtual Experiences* 79–97).

Virtual Harlem has also been used in courses not only in the United States (Columbia University and the University of Arizona) but also in Europe (the Sorbonne in Paris and Växjö University in Sweden).[1] It is a central exhibit at the SciTech Hands-On Museum in Aurora, Illinois, and was the centerpiece of the Black Creativity exhibit at the Museum of Science and Industry in Chicago during February 2002. Moreover, Virtual Harlem has inspired a variety of VR projects.

The Virtual Bronzeville Project

Virtual Harlem stirred considerable interest in the African American community after Carter received an award for creating it at the 2001 Marketing Opportunities in Business and Entertainment convention in Chicago. As I mentioned earlier, a group from Bronzeville, headed by Carol Adams, the director of the Center for Inner City Studies,

Figure 2.   A QuickTime version of Virtual Harlem

requested a showing of Virtual Harlem. After viewing Virtual Harlem, the group was determined to create a similar VR model for Bronzeville; this interest led to a series of discussions. The Virtual Bronzeville Group is motivated by the educational potential of projects like Virtual Harlem. Most of the group's members are actively involved in saving buildings of historical importance in Bronzeville because they do not want to lose vital connections with the storied past of their neighborhood. As I noted, many sites have already been torn down, and those that remain are in serious need of repair. As a consequence, the group is persuaded that Virtual Bronzeville would be an ideal way to preserve the neighborhood's past, which was an important chapter in African American cultural history.

Although in recent memory the Near South Side of Chicago has been associated with crime and violence, earlier in the century it was a thriving neighborhood and an African American cultural center. In "Chicago's Black-and-Tans," reprinted in *Chicago History*, the magazine of the Chicago Historical Society, from *Chicago Jazz: A Cultural History*. Joseph "King" Oliver, Louis Armstrong, and other famous jazz musicians performed in the cabarets in Bronzeville. The article includes a map of Bronzeville showing dozens of famous cabarets in the district from 1914 to 1928. The area was home not only to jazz musicians but also to figures like Andrew Dorsey, sometimes called the father of gospel music.

Music was not the only cultural outlet available in Bronzeville between the world wars; the South Side Community Art Center schooled innumerable aspiring painters and poets in their crafts (see Figure 3). Many African American artists, poets, and novelists lived in Bronzeville. The *Defender*, the nation's most widely read African American newspaper, was founded in 1905 by Robert Abbott, Chicago's first black millionaire. Its offices have been located in Bronzeville for a century. From its inception it has been a newspaper that has persistently fought racial prejudice. Its wide circulation in the South, where employees of Pullman's railroads carried it, is usually cited as a key factor in the migration of African Americans to Chicago. Langston Hughes was, for a time, one of its journalists.

Many of the children growing up in the neighborhood know nothing of this history. To preserve Bronzeville's past, residents like Harold Lucas have designed curricula to

Figure 3. VR replica of the South Side Community Art Center

teach them about Bronzeville. It begins before the waves of African American immigrants to Chicago and continues up to the present day gentrification. Lucas's intention is to give Bronzeville residents pride in the place in which they live.

Thanks to the help of the directors of UIC's Electronic Visualization Lab (EVL), Tom Defanti and Dan Sandin, programmers working there have started building Virtual Bronzeville. The model will differ from Virtual Harlem, owing to changes in technology and lessons learned. However, at the heart of the project is the shared intention to develop a collaborative learning environment in which students who wish to study Bronzeville can both learn from what has been done and contribute to its construction.

To students of jazz and blues, it will not be a surprise that a group at the Sorbonne in Paris who have seen Virtual Harlem wish to construct a Virtual Montmartre. Many of the famous African American musicians, writers, and artists who lived and worked in Harlem or Bronzeville, often in both neighborhoods, also lived in Montmartre. Carter, who has frequently lectured at the Sorbonne on the Harlem Renaissance, was invited to develop Virtual Montmartre.

What makes Virtual Harlem, Virtual Bronzeville, and Virtual Montmartre important to teachers of cultural history is that these VR simulations are reproductions of historical settings in which students can experience, however virtually, sights that no longer exist. They are resonant examples of VR as an instructional technology that makes history come alive. The effect, particularly of CAVEs, is like going inside a film and becoming an actor in it. Persons exposed to such effects experience the past in a sensuous way. For example, one of the VR booths at DisneyQuest had visitors take a virtual ride on a roller coaster. Those who did usually experienced physiological effects much like those produced by actual roller coaster rides. VR is capable of providing the impossible—a living experience of the past.[2] VR scenarios provide experiences that might otherwise be unavailable to learners. Because it is not possible for us to experience the past, virtual histories are helpful in giving students a "holistic" sense of what it would be like to have been in the past. VR is, in this respect, a kind of time machine.

Collaborative Learning Environments

Virtual Harlem is a learning environment in which participants experience a dramatic, visual history centered in Harlem, New York. By modeling its historical context as a dynamic system of social, cultural, political, and economic relations, Virtual Harlem enables a subject matter like the Harlem Renaissance to be studied in new ways. Unlike a conventional classroom in which the subject matter being studied is available to students as information in textbooks, on blackboards, or in slides projected on a screen, Virtual Harlem is an experience of the past. Students enter a cityscape that can be experienced, albeit virtually, as if they were tourists visiting Harlem via a time machine.

Techniques of computer simulation and visualization now make it possible to present historical events in virtual space and time. Though scholars disagree on this point, Harlem is widely thought to be the most important historical location for events that comprise what literary historians refer to as "the Harlem Renaissance." Although the project has not yet developed to this point, its organizers plan to show the historical changes in Harlem from the 1920s to the mid-1930s.

Building Virtual Harlem requires historical research and promotes learning by modeling, which is one of the most important facets of Virtual Harlem as a learning network. Recent developments in educational theory confirm that building computer models of the subjects being studied is for many students an effective way of learning about them. While such techniques have been used for years in Architecture and Fine Arts departments, computer modeling did not spread rapidly into other subject areas until computers became widely used educational tools. Computer-assisted drawing (CAD) programs have been employed in engineering studies for decades. The development of geographic information systems (GIS) software, to take a more recent example, is now used extensively in urban planning. Students in cognitive psychology use computer-generated models of cognition as a featured aspect of their methodology. Software such as Stella or Inspiration enables children in elementary or primary school to study scientific subjects by making computer models of them. Modeling has entered humanistic study. Shakespeare studies, for example, include models of the Globe Theater and programs that allow students to stage plays.

Virtual Harlem configures a "neighborhood" comprised of many elements: buildings, people, cars, events, communications, markets, and other phenomena.[3] Taken together, they comprise a dynamic system of relations. People live in buildings, pay rent, buy goods, make decisions, respond to injunctions, talk, sing, dance, drive, and involve themselves in multifarious relations with the other elements in the immediate environment. Computer models allow for the computation of a variety of possible systemic relations and provide a way of understanding the historical period by configuring it (see section titled "Learning by Configuring").

Virtual Harlem is a dramatic presentation of the history of the Harlem Renaissance. Scripts of everyday life are built into the presentation to dramatize the historical events (Sosnoski, *Virtual Experiences* 167–97). From a technical standpoint, VR interactivity is a promising area of research with untapped applicability. At present, students can interact minimally with figures that "live" in Virtual Harlem and whose character and behavior are as historically accurate as we can make it. Though such experiences are fictive by definition, the dramatizations are governed by scholarly efforts to interpret what it felt like to live in Harlem during the 1930s and to encounter the many artists who worked there.

While admittedly an unconventional form of history telling whose historiography has yet to be theorized, every effort is being made in ASCEND projects to give students an experience of the past that matches scholars' interpretations of it. The governing genre in this endeavor is history, not fiction or even historical fiction. The fictive elements are inevitably added to the existing documentation to fill in gaps in our historical knowledge, much as in the case of documentary films. Whereas it is possible to write sentences such as "residents of Harlem could purchase *The Crisis* at a local newsstand," a dramatization of that event requires a specific figure to approach the newsstand and ask for a copy of *The Crisis* (Sosnoski and Carter, *Virtual Experiences* 177–83). Since we do not have photographs of that event or a recording of what was said, that figure in Virtual Harlem cannot represent an actual person who lived in Harlem at the time. Yet to dramatize the historical generalization ("residents purchased *The Crisis* at local newsstands") does not entail the genre of fiction. The stories told in Virtual Harlem are governed by historical constraints.

From another perspective, Virtual Harlem is a project in "urban archeology." We have plotted out the surface of historical Harlem and drawn a map of its topography. At various locations on the map, we have dug deeper into its history to obtain a closer look at the development of that site. For example, whereas some buildings are no more than facades to mark the space they occupied at a particular moment in history, others can be explored in much more depth of detail. What the researchers unearth about a particular place is then recreated virtually. As a representation of a "neighborhood" in a city, the Virtual Harlem project can be extended to other neighborhoods in New York City. As a representation of a city, the Virtual Harlem project can be extended to other cities and their neighborhoods.

Learning by Configuring

Up to this point, I have focused on the learning perspectives one can take in using the various features of Virtual Harlem as an instructional tool. Let's now view Virtual Harlem from the perspective of the learners. Since there are several "learning pathways" learners can take in Virtual Harlem's environment, there are several possible outcomes available to them. I will focus on two general outcomes: for visitors and for builders.

The Virtual Harlem project is based on two hypotheses. First, visitors to Virtual Harlem who "interact" with figures of the Harlem Renaissance frequently during the course of a semester are likely to undergo a "deep learning" (Nokes and Ohlsson) experience during which they reconfigure their views of African American culture. Second, persons engaged in building a model of Virtual Harlem are likely to undergo a "deep learning" experience during which they reconfigure their views of learning. These educational experiences are not available in conventional classrooms.[4]

As a "configuration" of the Harlem Renaissance, Virtual Harlem allows visitors to reimagine Harlem as one of the seminal locations in the development of African American culture.[5] Because it provides a virtual experience of the Harlem Renaissance, it has the capacity to configure or reconfigure this period in visitors' memories. In other words, it has the capacity to introduce images of behavior or "scripts" (Schank and Cleary, *Scripts, Plans, Goals, and Understanding*; Schank and Cleary, *Memory Revisited*) into a visitor's worldview (Cobern). The modeling techniques used in this approach, which are present in the Virtual Harlem Project, can lead to a form of "deep learning," which I refer to as "reconfiguring" (counterstereotyping as a form of cognitive reframing). Most visitors alter their view of African American culture to varying degrees.

Stellan Ohlsson, a cognitive psychologist at UIC who has studied VR learning environments, argues that ideas that are fundamental to knowledge domains are acquired through "deep learning."[6] On his Web site, he writes: "Unlike other types of knowledge, fundamental ideas cannot be acquired through discourse or concrete experience, because those ideas are the very tools by which the mind interprets both discourse and experience." Such ideas are acquired through a process he terms, "deep learning," during which the cognitive frameworks (i.e., the abstract general frames or concepts) persons use to conceptualize their experience in particular knowledge domains undergo a transformation. According to Ohlsson's Web site, "New fundamental ideas are acquired by instantiating an abstract schema in a novel way; the new instantiation gradually assimilates pieces of the relevant domain, until it has effectively become the new center of that domain. Abstract schemas, in turn, are generated by combining and transforming prior schemas." His research points to the following observation: reading about a theory has little impact on students unless they have acquired "an abstract schema of the fundamental concept *in another domain*."

Ohlsson's research includes experiments with VR scenarios that provide "another domain" through which students can experience "fundamental concepts" before they learn them in the context of the knowledge domain with which experts associate them. UIC's EVL (http://www.evl.uic.edu) is a world leader in VR technology. Two of its faculty, Andy Johnson and Jason Leigh, have been instrumental in the development of Virtual Harlem. They have also worked closely with Ohlsson who writes on his Web site:

> New technologies for presenting interactive 3-dimensional worlds have been developed at UIC's Electronic Visualization Laboratory (EVL). This technology is a means for presenting students with alternative experiences that contrast with everyday experience in educationally relevant ways. The objective of ["the Round Earth"] project [was] to explore the potential of virtual reality to support deep learning. During the fall of '97 and spring of '98, a pilot project [used] virtual reality to teach young children that the Earth is round, a concept that prior research has shown is difficult to grasp. Future applications of virtual reality will focus on more complex learning targets.

Working from the assumption that the "deep learning" process is often highly "analogical," I extended Ohlsson's research into the ways in which persons create maps of their personal worlds—the mental constructions of space that persons use to locate their experiences. Since Virtual Harlem is a representation of African American culture, it matches that part of a person's worldview that comprises Harlem. One hypothesis governing the project is that sectors of personal worldviews parallel knowledge domains. Unlike knowledge domains, which "give order to abstract concepts," the components of worldviews provide general narratives, "configurations," that shape personal experiences as a "way of worldmaking" (Goodman). These configurations (i.e., narratives of personal interactions) make up worldviews that are situated, cognitively, in a personal map of the world.

Configurations undergo a process of transformation similar to the one Ohlsson describes with respect to "fundamental ideas." Because configurations are the stories that are inscribed in personal maps of the world, they can be said to constitute a "cultural domain" parallel to a knowledge domain. Such domains are often shared—in the case of knowledge domains by the members of a discipline, and in the case of cultural domains by the members of a culture. Theoretically, then, it is possible that a person may replace a stereotypic configuration of African American culture with one that reflects its history more accurately. Persons may enjoy transformative learning experiences during which their mental construction of a city neighborhood is transformed (reconfigured) as the result of virtual experiences of its history.

Worldviews, in this theory, are formed by a thought process called "configuring." In this process, the memory stores experiences through cognitive scripts (i.e., stories), which are very general since they are abstracted from experiences. These narrative abstractions are sometimes referred to as "memory organization packets" (MOPs), "scenes," or "scripts."[8] On the basis of these scripts, persons draw inferences about human behavior. For example, when a person encounters a man who is wearing an Armani suit, he or she might draw the inference that this person is "wealthy." This is not a logical inference but an analogical one based on the "model" of "wealthy men" in his or her map of the world, usually derived from the media rather than actual encounters with wealthy men. The "model" of a rich man in a person's map of the world is usually embedded in a set of "scripts" that portray the "behavior" of the rich and famous. If you ask a person who has never actually experienced wealth what rich men do, answers to the question by that person are drawn ("scripted") from the "configurations" in his or her map of the world, often mediated by TV and films.

Another example would be the configurations created in the American populace by the media coverage of events such as hostage crises and the Gulf War.[9] When a culturally dominant group challenges the configurations shared by a minority group, those configurations are usually (somewhat condescendingly) called "myths." Behind every prejudice is a set of myths populated by scapegoats and stereotypes. The mass media often depends upon them. African Americans are subject to various myths about their cultural behavior. Shows like *I Spy*, *The Cosby Show*, or even the recent more realistic TV shows about African Americans no doubt have some effects on reconfiguring the African American experience for many Americans. However, such *fictional* shows can only have a limited impact, and whatever impact they may have is difficult to study.

As I emphasized earlier, Virtual Harlem is a construction of the *historical* Harlem, New York, during the height of the period known as the Harlem Renaissance as a VR scenario. Recall that the aim of the project is to allow visitors to experience the Harlem Renaissance virtually by playing roles in the stories of the figures who lived in Harlem at the time. We anticipate that these virtual experiences of the Harlem Renaissance will result in transformative learning—reconfigurations of African American culture. This seems likely since the phenomenon of "configuring" entails identification with a figure (i.e., recognition of resemblances between virtual experiences and personal experiences). Such identifications are similar to the ones readers of novels or viewers of films experience, and they usually produce empathy for certain characters. In extended experiences of Virtual Harlem, users/visitors are most likely to understand the historical scripts by analogy to their personal experiences. This allows for the incorporation of the virtual experiences into the cognitive map of their world, potentially altering or replacing prior stereotypical scripts of African American culture.[10] This cognitive reconfiguration is dependent upon the *bodily* experience of being immersed in the historical virtual environment because it depends upon an empathetic emotional response to the virtual experience.

What visitors to Virtual Harlem may learn differs in substantial ways from what the builders of Virtual Harlem learn. Virtual Harlem is an instructional technology that integrates various technologies to form a networked learning environment. The persons engaged in building the Virtual Harlem environment have the opportunity to learn how to deploy the technologies it employs. At the minimal level, builders learn how to use the various software or hardware involved. This can be as simple as learning how to send a file as an attachment, or it can be as complex as learning Yrgasil, the version of C++ used in the project. Most students become acquainted with 3D graphical software or with video conferencing software and hardware as a direct result of their work on the project. No student is required to learn any software in order to participate and may help build the environment by conducting traditional research in the library in collaboration with someone who is more adept with the requisite technology.

The Intersecting Learning Paths in VR CLNs

It may be helpful in understanding the structure of a CLN to introduce the idea of a "learning pathway."[11] Sometimes discussions of learning seem to imply that learning takes place *only* in a classroom. As we all know, this is far from the case. We often learn more outside the classroom than in it about a particular subject. The Virtual Harlem project requires students to go outside of their classrooms and visit other sites. Several learning paths (i.e., movements from site to site) typically occur in the Virtual Harlem project. Let me begin with a student taking a course in the Harlem Renaissance.

If we think of a learning pathway as a journey of discovery, then we might designate a conventional English literature classroom as the site of departure. A student may be assigned to read a work introducing the Harlem Renaissance as a period of literary history from his or her textbook. This would lead him or her home or to the library where the reading might take place before the student returns to the classroom. This "trip" would be repeated many times in the learning path. At some point, that student may go to the immersion CAVE instead of the classroom and experience Virtual Harlem and then perhaps take a trip to a computer center to record her response to it. If he or she decided to take on, as a term project, a research endeavor that culminated in adding some information about a building that was yet to be built in Virtual Harlem, this would necessitate a trip to the library and to a computer lab with Internet access in search of photos and accounts of, say, the Dark Tower (an important salon). This decision would likely lead to the preparation of a research paper not only submitted to the instructor but also presented for "publication" in Virtual Harlem. Since a literature student or instructor would probably not be able to build a 3D Dark Tower into the VR scenario, the "publication" of the Dark Tower would be handled by someone who could.

The learning path of the literature student, in all likelihood, would have to intersect with the correlative learning paths of several other students before the Dark Tower could be added to Virtual Harlem. These intersections would "deepen" the learning experience to the extent that the students and instructors involved were in dialogue with each other. Without going into the same detail, I hope it is easy to imagine an engineering student beginning his learning path in a class on C++ and, at some later point, taking on the project of constructing in 3D code the image of the Dark Tower as his term project. A third student in fine arts might start out in a computer graphics class in the School of Art and Design and follow a learning pathway adding an aesthetic dimension to the design that eventually intersected with the first two students'. A women's studies student might intersect with the group in an effort to portray the women in the Dark Tower setting accurately. If they collaborated, they would have to take perspectives they would not otherwise have considered into account and would thereby enrich their learning experiences.

These learning paths would also inevitably intersect with those of students from other universities who were researching the Dark Tower or some of the persons that were its habitual guests. If some of the learning paths did not intersect smoothly but, instead, clashed and contradicted each other, the learning would take a steeper turn as students would then be forced to encounter perspectives that clash with their own, a likely scenario as more universities from abroad join the network. In this case, learning would involve negotiating the differences in perspectives.

The Feasibility of VR as an Instructional Technology

In recent years, there has been an explosion of VR programming tools used to create interactive video games that, as I have already noted, can be used to develop VR educational applications as well. The products of these tools are delivered to users on a variety of operating systems, ranging from those designed for desktop and laptop computers to more powerful systems employed in CAVEs. The most powerful of the VR environments is the CAVE, aptly named since it is a room with screens on every side of the persons who enter it. CAVEs are "immersive" VR applications because the persons who experience them are literally immersed in a scenario created by as many as six walls onto which images are projected.

Though the cost of programming VR scenarios is at this time prohibitive unless grants are obtained and though the learning curves for VR technologies are steep for persons without technical backgrounds, a number of universities have been developing VR learning environments in their Computer Science and Engineering departments. As the technology advances, we can expect the costs and programmability constraints to be lowered. Efforts to cut the costs of VR installations are already well under way at the EVL, where a one-walled 3D VR setup, called a GeoWall (see Figure 4), has been designed and made available on a wide scale. The projects mentioned earlier are designed for GeoWalls. The SciTech Hands-On Museum has such an installation and Xperiment Hus in Vaxjo, Sweden is soon to build one. The Virtual Bronzeville and Virtual Montmartre projects will be displayed on GeoWalls at the Center for Inner City Studies (CICS) and the Supreme Life Building Tourism Center in Bronzeville, and the Sorbonne in Paris. Moreover, these projects can be shown in a variety of ways with modest changes in the programming code. Virtual Harlem, for example, can be presented in a video game format as well as a 3D graphical display on any PC or Mac or sent across the Internet as a QuickTime video.

Though the development of VR instructional technology is in its inception, given the cultural impact of video game technology and the popularity of sites such as *Second Life*, it does not seem unrealistic to predict that VR will be a significant teaching tool in the next decade.

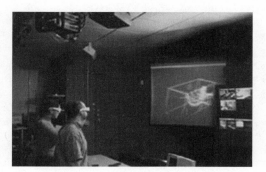

Figure 4.   A GeoWall

Notes

As I return to this chapter some months after I wrote it, I have to note that the Virtual Bronzeville project has been delayed by local political issues. The members of the group who initially contacted me to help design this project were not able to agree on central issues of emphasis, audience, and the interpretation of the site's history.

1. The universities mentioned belong to ASCEND (Arts and Science Collaborative Network Development). See Sosnoski and Carter. When I use "we" in the discussion of ASCEND projects, I am referring to the project organizers.

2. The concept of "null experiences"—experiences that persons have not yet had or cannot have—is not only pertinent to history but also to intercultural and gender communication because men cannot experience childbirth and persons who are white cannot experience the racial prejudice directed against African Americans. See "Configuring African American Culture as Virtual Experiences of History" in *Teaching History and Configuring Virtual Worlds*.

3. See Sosnoski, "Alternative Cultures" and "Explaining, Justifying, and Configuring."

4. Configuring is an "experience transfer" by which persons project experiences they have had by analogy onto other persons or situations or introject by analogy other persons' experiences into their memories. In a typical knowledge transfer (Nokes 2004), a familiar concept or pattern is mapped onto an unfamiliar one. In an experience transfer, a familiar experience is mapped onto an unfamiliar one to achieve a particular outcome.

5. See Sosnoski, "Alternative Cultures" and "Explaining, Justifying, and Configuring."

6. See "Description of Deep Learning" by Stellan Ohlsson. http://www.uic.edu/depts/psch/ohlson-1.html

7. See http://www.uic.edu/depts/psch/ohlson-1.html. Accessed 2002.

8. See Schank, *Dynamic Memory*. Schank also refers to MOPs as "scenes" or "scripts."

9. See Said, *Covering Islam*.

10. We have yet to gather data on this hypothesis but have developed several grant applications to acquire funds to do so.

11. This section is taken from an earlier essay of mine, "Will New Technologies Impair the Critical and Imaginative Capabilities of Students?: Virtual Harlem, an Experiment in Learning Environments" (in Sosnoski and Carter, *Virtual Experiences*, 115–32).

Works Cited

Cobern, William W. *Everyday Thoughts about Nature: A Worldview Investigation of Important Concepts Students Use to Make Sense of Nature with Specific Attention of Science.* Dordrecht: Kluwer Academic Publishers, 2000.

Drake, St. Clair, and Horace H. Cayton. *Black Metropolis: A Study of Negro Life in a Northern City.* Chicago: University of Chicago Press, 1993.

Fauconnier, Gilles. *Mappings in Thought and Language.* Cambridge: Cambridge University Press, 1999.

Goodman, Nelson. *Ways of Worldmaking.* Cambridge, Mass.: Hackett, 1978.

Johnson, Andrew, Jason Leigh, James Sosnoski, Bryan Carter, and Steve Jones. "Virtual Harlem." *IEEE Computer Graphics, Art History, and Archaeology* Sept./Oct. 2002: 1–8.

Kenney, William. "Chicago's Black-and-Tans." *Chicago History* 26.3 (1997): 5–31.

Nokes, Timothy. "Investigating multiple mechanisms of knowledge transfer." Accessed 2004. Power Point Presentation.

Nokes, Timothy, and Stellan Ohlsson. "How is Abstract, Generative Knowledge Acquired? A Comparison of Three Learning Scenarios." *The Twenty-Third Annual Conference of the Cognitive Science Society.* Eds. J. D. Moore and K. Stenning. Hillsdale, N.J.: Lawrence Erlbaum, 2001. 710–15.

Park, K., J. Leigh, A. Johnson, B. Carter, J. Brody, and J. Sosnoski. "Distance Learning Classroom using Virtual Harlem." *Proceedings of the Seventh International Conference on Virtual Systems and Multimedia.* Berkeley, Calif.: VSMM. 489–98.

Said, Edward. *Covering Islam: How the Media and the Experts Determine How We See the Rest of the World.* New York: Pantheon, 1981.

Schank, Roger C. *Dynamic Memory: A Theory of Reminding and Learning in Computers and People.*
    Cambridge: Cambridge University Press, 1982.

Schank, Roger C., and Chip Cleary. *Engines for Education.* Hillsdale, N.J.: Lawrence Erlbaum, 1995.

———. *Memory Revisited.* Cambridge: Cambridge University Press, 1999.

———. *Scripts, Plans, Goals, and Understanding: An Inquiry into Human Knowledge Structures.* Hillsdale, N.J.:
    Lawrence Erlbaum, 1997.

———. *Virtual Learning.* New York: McGraw Hill, 1997.

Sosnoski, J., Steve Jones, Bryan Carter, Ronen Mir, Ken McAllister, and Ryan Moeller. "Virtual Reality as a
    Learning Environment: The Ascend Group." *International Handbook of Education Series.* Ed. Jason Nolan,
    Peter Trifonas, and Joel Weiss. Toronto: Kluwer, 2003.

Sosnoski, James J. "Alternative Cultures." *Modern Skeletons in Postmodern Closets: A Cultural Studies
    Alternative.* Charlottesville: University Press of Virginia, 1995.

———. "Configuring as a Mode of Rhetorical Analysis." *Doing Internet Research.* Ed. Steve Jones. London:
    Sage, 1999.

———. "Explaining, Justifying, and Configuring." *Modern Skeletons in Postmodern Closets: A Cultural Studies
    Alternative.* Charlottesville: University Press of Virginia, 1995.

———. "Prologue on Configuring." *Token Professionals and Master Critics: A Critique of Orthodoxy in Literary
    Studies.* Albany: State University of New York Press, 1994.

Sosnoski, James J., and Bryan Carter, eds. *Virtual Experiences of the Harlem Renaissance: The Virtual Harlem
    Project.* Spec. Issue of *Works and Days* 19.1–2 (Spring/Fall 2001).

Sosnoski, J. J., P. Harkin, & B. Carter, eds. 2006. *Configuring History: Teaching the Harlem Renaissance
    Through Virtual Reality Cityscapes.* New York: Peter Lang.

———. *Teaching History and Configuring Virtual Worlds: Virtual Harlem and the VERITAS Studies.* Ed. Steve
    Jones. Digital Formations. New York: Peter Lang. Forthcoming.

10

Digital Provocations and Applied Aesthetics:
Projects in Speculative Computing

Johanna Drucker

The *Speculative Computing Lab* was founded in 2000 with the goal of creating projects that would address what we saw as a serious crisis in the humanities. Briefly stated, that crisis is the striking disconnect between the culture of media in which students are immersed in their daily lives and the traditional approach to knowledge production and preservation that is at the heart of the classroom environment at every level of education. How will the basic mission of the humanities—to create and preserve culture—survive if the arenas that have traditionally maintained the viability of humanist values are rendered obsolete? Something has to change—radically and soon—if our cultural legacy is to remain within the horizon of a younger generation's view.

SpecLab projects were built on the foundation established by work in digital humanities at the University of Virginia, and Jerome McGann, my co-founder, was among the early adopters and pioneers of work done at the Institute for Advanced Technology in the Humanities (IATH). That work was reaching a critical stage of maturity by 1999, when I arrived at the university. IATH's development had been coincident with the emergence of the World Wide Web. Under the intellectual leadership of John Unsworth, its mission had been defined to maximize the potential of the Internet as an integral part of the design of digital humanities projects. Digital humanities (in its longer history, traced to Father Busa and his linguistics corpus; in its shorter history, at UVA and elsewhere) had been conceived in the traditions that derive from the archive and library.

But *SpecLab*'s mission was conceived as that of making a bridge between these traditional modes of scholarship, which often remain traditional in conception though they exist in digital form, and an emerging culture raised on games and instant access networks of socialized information exchange. The novelty of *SpecLab*'s projects within the digital humanities community comes from our emphasis on making tools for interpretation. Rather than creating discipline-specific, content-rich archives or databases, we have been creating environments for analysis that try to expose assumptions and call attention to subjectivity as an integral part of all interpretative activity and experience. We were initially interested in games and toys and in using visualization and aesthetic provocation as central parts of our designs. Our focus on designing interfaces to emphasize *visual* forms of knowledge production and providing analyses made our work distinctive within the community of text-based humanists. Though the creative, envisioning phase of SpecLab is over, the implementation of projects at *SpecLab* continues to evolve within the larger context of challenges for future scholarly and learning environments and their capacity to link traditional analog and evolving digital forms of knowledge production. We do this by integrating graphical display, game-type approaches, and playful, socially networked motivation structures for the serious pursuit of humanist work. Our commitment to aesthetic provocation is grounded in the belief that stimulating the imagination by offering intriguing and sometimes unfamiliar ideas in seductive form is crucial to opening the doors of perception.

Current Contexts and Challenges

In the last decade, the transformation of media use—including the use of personal electronic devices, cell phones, and other portable, "small" technology for communication and information access—has been extremely rapid. The full effects of these combined changes on the conditions and culture of education have yet to be fully assessed, but we know for sure that the students who are entering our classrooms (whether in K-12, college, university, or at the research level), have their expectations about daily patterns of information acquisition and exchange shaped in an environment with an unprecedented level of media saturation and access. The consumption of mediated stimulation is almost instantaneous, nearly continuous, and, it seems (if cell phone use is any indication), highly addictive. Just as the Web was pushed by the messaging capabilities of e-mail, so the media environment of current (at least first world) culture is driven by immediate gratification of the impulse to be connected. Chatting, messaging, searching, surfing, listening, downloading, archiving, playing, trading, buying, and selling promote an "all media all the time" sensibility. This unparalleled growth in connectivity signals a change for education, and tasks, motivation, and structures of learning will change dramatically in the decade ahead.

What part will educators and humanists play in shaping the future learning environment? The future learning environment is a moving target. Aiming toward the horizon of the imaginable allows us to keep just ahead of the recent past. The task of designing new technology is not so much an engineering challenge as a conceptual one. Ways of thinking about learning are emerging before our eyes in protean forms rarely stable enough to be fully defined. Now modeling a future learning environment on the format of a multiplayer online game seems to make complete sense. After all, that *is* what learning is and has been—an asynchronous, geographically distributed network of exchanges within a community of users who contribute, vet, create, and break consensus models of understanding and knowledge. As in all instances of media innovation, new media give us an explicit understanding of functionalities in older media that were implicit in traditional forms; for example, the electronic spaces of reading tell us much about the dynamic functions and features of paper-based books.

Some basic assumptions on which electronic instruments for research and pedagogy might be built are quite clear. Motivation is highest within a social context of recognized achievement. Wiki-type consensual knowledge production encourages active learning and participation. Collaborative research and pedagogy can be structured into game-like play. Information and learning tasks will be chunked for delivery in "bites" that can be accessed in portable, mobile, devices that serve as portals and access nodes with maximum display and limited processing capability. Knowledge and information assessment that is vetted (one can imagined "branded" information from "reliable" sources coming into the market) through a system of peer review and filtering could be combined with wiki-type knowledge production. The use of structured data for analysis of patterns, discourse formation, and idea formation requires innovative research questions to maximize the capabilities of research and teaching. Visualization will be essential for data analysis and interpretation. All learning, knowledge, and expression of information is subjective. Socially produced, consensual warranting of standardized, authoritative knowledge (including traditional peer review and professional credentialing) can be effectively administered and usefully advanced through digital environments. We can even describe all types of artifacts for knowledge production (books, texts, images, music) whether in digital or analog format, not as static objects but as probabilistic fields. This is an unfamiliar idea until you try it, and realize that any reading or experience of a work or text produces that work anew. That act of reading could be construed as an "intervention" in the "probabilistic" field of a text. The advantage of thinking about texts or knowledge environments in this dynamic way is that it sets the foundation for thinking about how these environments are shaped and formed through ongoing, emergent processes. Group participation, strategic short-burst, high-stimulation engagement and response, and sustained identification with virtual spaces of participation are the mind-drugs of the current future.

Projects

Of the projects we began at *SpecLab*, several developed to the point of fully usable functionality. Others remained in beta, exist as proof-of-concept prototypes, or are concepts still under development. The process of design creation and production of Ivanhoe, Temporal Modeling, Juxta (a collating tool for bibliographical studies), Collex (an online collection and presentation tool), Subjective Meteorology (a visualization scheme for emotional experience), and the Patacritical Demon (a tool for reflection on processes of interpretation) has been a striking educational experience. Ivanhoe and Temporal Modeling will be the focus of discussion here. The "game" of developing these projects within the collaborative and iterative process has revealed the addictive character of participatory, distributed, aggregated learning. In short, SpecLab has embodied the principles it seeks to design in its projects. But we have much to learn still about how to instrumentalize some of these insights within a productive, task-focused environment for learning. And by "learning" it should by now be clear that I am referring to the tasks of research and writing across the broadest spectrum of populations, from students to teachers, credentialed professionals to amateur enthusiasts, children to adults, and casual to focused users of information.

    SpecLab's projects have a deliberate intellectual affiliation with the work of French poet and philosopher Alfred Jarry. They each have a touch of the ludic in their conception (the demon most of all) through their association with pataphysics, or the science of exceptions, as an exercise in imaginative and generative critical thought. The lab for software development is named ARP—applied research in patacriticism—to signal that sensibility. But the roots of these projects are also deeply grounded in textual studies, bibliographical method, graphesis (visual knowledge production), critical theory, systems theory, philosophy, and semiotics (in its Peircian as well as classic structuralist forms). We have flirted with game theory and had occasional formal sessions on "the ludic" (i.e., playful) as a mode of serious thought. The aesthetic features of *SpecLab* designs embody intellectual and critical issues central to a theory of performative interpretation and also aim to provide effective cognitive means for engagement with critical work. Meant for scholarly purposes as well as pedagogical ones, these projects extend the understanding of subjective interpretation, which is the very foundation of humanistic inquiry.

The two earliest and most developed projects to date, Ivanhoe and Temporal Modeling, were both responses to the same question: If a digital environment could be designed to increase self-consciousness about the nature of subjective interpretation in humanities work, what would it look like and how would it function? Temporal Modeling was conceived as a visual composition space for representation and analysis of the complex temporal relations that exist within and among humanities documents. Ivanhoe is more squarely located within textual studies as an extension of theoretical investigations of textuality in its material, graphical, literary, critical, semiotic, social, and cultural dimensions. Both are designed to emphasize the performative nature of interpretation, drawing on principles that are the legacy of deconstruction, critical theory, and cognitive science.

These projects had different original impulses. Temporal Modeling was created in response to an existing project, and its parameters were conceived almost at the outset. Ivanhoe emerged through an iterative process, first in a spontaneous exchange that involved many research sessions and discussions with a wide range of participants. Temporal Modeling exists as a working prototype, and Ivanhoe is now a functional piece of software. Temporal Modeling has had a concrete, tangible outcome with a usable tool as its goal and result. Ivanhoe spawned many ongoing conversations about digital and traditional textuality and interpretation. It is also a project to design an electronic environment for digital and traditional textuality and interpretation. Many prototypes and versions were instantiated along the way in paper, e-mail, and electronic forms.

Temporal Modeling

Temporal Modeling was conceived early in 2000 after I saw a demonstration by John David Miller of Intel Corporation. Miller was working with John Maeda at MIT to create an electronic space for the chronological display of documents. Their project attempted to maximize the use of screen real estate, provide flexibility of organization, and provide the user with an intuitive interface. I showed the prototype to some of my colleagues and suggested that we come up with alternatives to some of its basic assumptions and design features. Developed with the assistance of Bethany Nowviskie, Temporal Modeling was created as a working composition (or "play") space that realizes our (SpecLab's) original vision on conceptual and technical grounds. The "provocative" use of aesthetics in this work is twofold. Visual means are used in the composition space as a primary method of producing interpretation and for purposes of analysis. This would hardly seem radical to visual artists, but for humanists, such ideas—and the tools to explore them—are unusual.

Because of the technically constrained nature of the visual tool set, the composition space is able to produce XML (extensible mark-up language), the structured data that is the standard format for humanities archives and electronic document management systems. In Temporal Modeling, the aesthetic image "provokes" the structured data as a simultaneous response (see Figure 1).

Temporal Modeling offers an alternative to standard timelines developed largely for empirical studies. Standard timelines are based on three assumptions. The first assumption is that time is unilinear (the time arrow is conceived to always go in one direction, from past to future, on an unalterable and single path). The second assumption is that time is homogeneous (time is a "given" with a standard metric that is uniform for all time, generally conceived as seconds, minutes, hours, days, etc.). Finally, it is assumed that time is continuous (all points along a line can be accounted for in the same system and all are equally available for analysis and representation). These ways of graphing time are effective for recording presumably "empirical" data based on strictly quantitative, highly rationalized methods of information gathering and display. According to these premises, the visual display of this quantitative information (to paraphrase the well-known information designer, Edward Tufte) aims to be as legible a presentation of the "actual" data as the designer can manage. Our system, by contrast, was based on the notion that the presentation creates the data, the interface is the primary input, and that the *modeling of temporal relations* is based on subjective experience. Temporal relations allow for a "multiple worlds" view of alternative narratives rather than a single, unilinear time arrow; for warped and often highly compressed or excruciatingly expanded, stretched experiences of time;

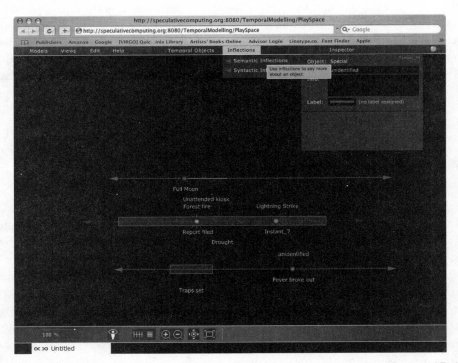

Figure 1.   The Temporal Modeling interface, designed at SpecLab by Johanna Drucker, Jim Allman, and Bethany Nowviskie with design assistance from Petra Michel

and for a noncontinuous, often broken or ruptured (even contradictory) account of temporal experience from subjective points of view (see Figure 2).

We wanted to create a visualization scheme that allowed for flexibility, ruptures, multiple narratives, modes of comparison at different scales, and relative rather than absolute temporality (for example, one might only be able to answer in relative terms the question of "when" something occured, such as "before Mama came home" or "after Dad went to work for the railroad"). Time in such documents is lived and experienced, not extrinsic and absolute.

Our modeling system includes a number of distinctive features: a "now-slider" (a device that could be moved through a display but always indicating a moment that was "now"), semantic "inflections," and syntactic "inflections." The now-slider anchors individual subjective perception within the system.

The now-slider's point of view allows multiple narratives and alternatives within a single display. Contradictory interpretations of events can be created in the composition space, each linked to a different now-slider identified with a particular individual. Comparative interpretations for a situation (showing who knew what when in particular historical or fictional circumstances—a political campaign, a mystery story, a battle) can be modelled to show contrasts in understanding linked to particular viewpoints as well as "omniscient" or "objective" ones. Subjectivity is marked as point of view and explicitly linked to the now-slider, but subjective interpretation is also enabled by the use of inflections, which are visual elements that can be used to characterize a temporal event either semantically or syntactically. Semantic attributes include degrees of importance or emotional character such as sadness or another quality or characteristic. Syntactic inflections involve a relational exchange such as foreshadowing, in which one event is given value

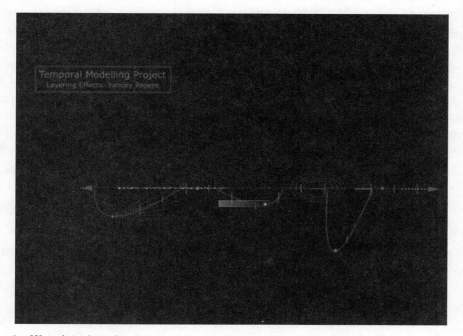

Figure 2.   Warped timelines: An experiment showing subjective inflection by warping timelines

through its connection to another. Inflections, the now-slider, and the capacity for alternate forward and backward (prospective and retrospective) branchings provide a graphical vocabulary suited to humanities interpretations, rather than simply seeming to "present" empirically grounded findings. The contradictions and complexities of fiction and historical record or the blend of public and private documents that characterize much humanities work in history, literature, anthropology, the arts, and cultural and media studies can be visualized in this system.

The technical capacity of the project is its ability to generate XML on the fly from the manipulation of design elements in the composition space. This is crucial to its connection to digital humanities. Rather than using a visual means as a final display, Temporal Modeling allows visual thinking to play a part in the early stages of what is called "content modeling"—the design of a data structure that can accommodate documents in an electronic environment.

The inclusion of subjectivity as an explicit design feature in this work proved particularly inspiring in some of our conversations about Ivanhoe. Designing Temporal Modeling involved several clear-cut steps. We began with research into the philosophical literature on time, temporality, and its conception.[1] We compiled a visual archive of ways of representing time and temporality. We surveyed the existing literature in logic, diagrams, and computational analysis to conform our vocabulary as much as possible to existing use. Finally, we created a distilled conceptual schema for the "primitives" of temporal modeling for the project and developed the visual design of the actual interface.

Ivanhoe

Ivanhoe emerged through a combination of game-play, design visualization, discussion, and prototypes. Jerome McGann and I began this project in the summer of 2000.[2] We initiated the project as a provocation in an exchange, reworking *Ivanhoe* as a critical act of interpretation, rewriting the story as a way to analyze it. So, the original challenge was to change the outcome of the Walter Scott novel so that one character, Rebecca, could be rescued from the fate dealt to her as the novel's conclusion; her decision to withdraw into convent life would be replaced, and she would be offered a chance for happiness in the arms of the ravishing seducer, Bois-Gilbert. Our initial exchange took place through e-mail. We wrote passages that were linked to locations specified in a mutually agreed-upon print edition of the work. The asynchronous character of the dialogue gave it a particular excitement—the unpredictability of timing, the thrill of imagining the recipient's amusement, the secret pleasure of upping the ante of the game through each increasingly daring "moves." Our contributions were largely text-based, though I did at least one drawing—showing Rebecca after several years in the company of her dark-spirited lover and far from England's repressive shores.

We quickly realized that the possibilities of the Ivanhoe project outstripped the simple means of play we had used in this first round, and we set about to formalize some of the fundamental principles. We realized that any "move" needed to be done as an explicit character of an explicitly characterized role. Each of us had been working implicitly from such a point of view—McGann as a Byronic type, myself as a sort of latter-day Colette-influenced feminist dedicated to the exploration of female liberation through romantic adventure.

The recognition that roles were, in fact, crucial to Ivanhoe put us on the road to serious game design—and made important conceptual links between pedagogy, research, and current cultural trends. But this realization also pushed a crucial theoretical issue to the fore. Making role-playing a self-conscious part of Ivanhoe would reinforce the located nature of interpretive acts, a tenet that has characterized critical studies for the last few decades. While recognition of subjectivity is a key feature of current critical thought, enacting that principle in pedagogy requires some imaginative work. The tendency is to slip into habits of familiar text-as-object and reader-as-subject approaches to reading. Ivanhoe deliberately blurs those boundaries.

Our emphasis on making subjective aspects of interpretation explicit pushed us into interface design. McGann and I crafted the design in a hand-drawn storyboard (Figure 3). We set about outlining a rule-set and came up with a Byzantine, elaborate list of rules whose chief attraction, it turned out, was the challenge of designing them. We flirted with a systematic approach to game design and tried two alternative models—one based on cooperation, one on competition—to be tested for research and classroom use.

In the initial interface, we imagined the game taking place from a "source text" as a series of "moves" composed in a "work space" that could make use of textual, musical, visual, or time-based media materials. Each move would be accompanied by an entry in a player journal or log that justified the way the "role" was creating an interpretation. The player journal would remain private unless a move was challenged, or until the game was over, or some other contingency forced the revelation of this reflective space. Reflecting on the reasons for a move was and is a central feature of Ivanhoe as a critical space, and the

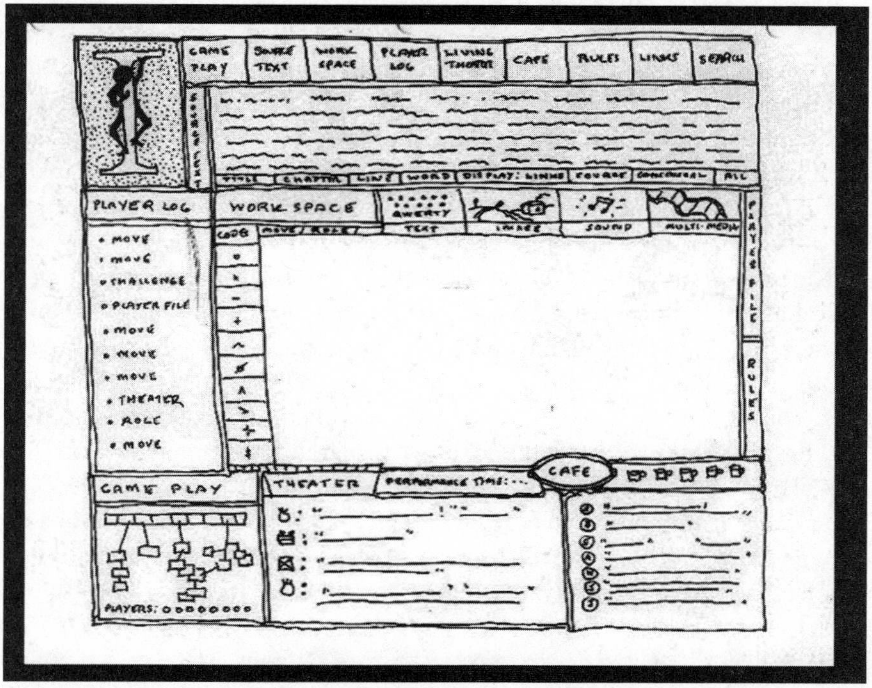

Figure 3. Early Ivanhoe: A hand-drawn mockup showing the functionalities we wanted to design into Ivanhoe software (by Johanna Drucker in consultation with Jerome McGann)

theoretical justification for this is to ensure that interpretation is practiced self-consciously rather than as a neutral or unmarked act. Creating awareness of the social production of texts is one of the pedagogical goals of Ivanhoe—to recognize the links between any individual instance of a work and the complicated field of versions, editions, source materials, and contributing elements that bring it into being. We term this extended sphere the "discourse field"—a zone that is unbounded by definition but capable of being invoked, particularly in electronic space, by a whole host of various documents.

Our original interface design included two spaces that supported the social nature of interpretation within research and pedagogical practice, the "living theater" and the "café". The theater space was designed to allow "roles" or "characters" to interact with each other in a real-time, MOO-like (online multi-player games created through texts and descriptions that were early virtual environments) feature while the café was designed to allow players to shed their roles and converse directly, commenting on the game, the moves, or the scholarly project underway. While Ivanhoe can be played in an entirely ludic spirit, we envision the use of the interface for collaborative and distributed scholarship as well. The performative spirit of Ivanhoe is grounded in deconstructionist theories of textuality. The Barthesian idea of a "text" as a field of play lurks in our insistence on critical practice as constitutive rather than as an act of revelation. The antitranscendent impulses of French theory, of Derridian *différance* as a principle of poststructuralist semiotics, are crucial to Ivanhoe's foundation—as are Bakhtinian notions of dialogue and dialectical thought. Designing an electronic environment that takes these now well-established approaches to critical thought into a new relationship with electronic archives and the host of bibliographical issues raised by digital technology is one of the challenges the Ivanhoe team has set for itself.

Self-consciousness about historical and cultural location, the social production of texts within an elaborate network of which the bibliographical evidence is one index, and the collaborative nature of scholarly work within a distributed community are the crucial features of Ivanhoe's conceptual design. Within this set of parameters, a number of other concerns have informed the design of the actual interface. How to use the electronic environment's specific graphical capabilities to call attention to such theoretical issues as the "non-self-identical" nature of a text, to discourage solipsism, to increase awareness of bibliographical features of paper-based as well as electronic documents, and to show the relational nature of interactions among players in the game were particular concerns (see Figure 4).

These concerns have pushed us to various experiments in prototype design and visualization. Several versions of Ivanhoe have been played (some more or less game-like). One was played with middle-school children using pen and paper. The rest have been played in e-mail, using a modified blogging program called Grey Matter, or in a prototype built by one member of the research team, Nathan Piazza, as a Web-based experimental model. The games have been played using Emily Brontë's *Wuthering Heights*, Madeleine L'Engle's *A Wrinkle in Time*, Mary Shelley's *Frankenstein*, Henry James's *The Turn of the Screw*, and William Blake's *The Four Zoas*. Experiments in interface and in visualizations generated from data produced through the game-play have put theoretical issues into direct dialogue with design considerations. Each iteration distills the project further, and

Figure 4.   Ivanhoe game space: The working application showing a game in process. Documents are distributed around the discourse field, and lines of influence map exchanges among players and between players and documents. Many other views of this data can be accessed from within the application, and the game is always displayed from someone's point of view.

its usefulness for pedagogy and research in the humanities can be articulated through both short-term technical and longer-term rhetorical goals.

Conceived at a particular moment in the history of higher education and textual studies, Ivanhoe extends the dialogue between traditional humanities and mass-media culture. Our conviction is that reading and literacy in the next generation of humanities scholarship will be intimately bound up with technological procedures. To date, the bulk of electronic instruments engaged for humanities purposes focus on library and information management systems for the administration and delivery of materials. While such uses of technology are essential, they are not sufficient. Creation of imaginative, provocative spaces for engagement will be required if the potential of digital technology is to be realized as an integral activity in humanities work. Much is at stake in conceiving of digital methodologies that extend, rather than curtail or circumscribe, humanistic thinking. If we are to move beyond instrumental management and statistical processing of text-based materials, then what issues will galvanize research at the intersection of digital technology and humanities research? And if the future of literary and humanities scholarship will depend upon electronic instruments, then on what foundations should these tools be established and what approaches to the study of print-based materials should factor into their design?

Humanities Games

The very acts of reading required by traditional humanities seem alien to many of the students in our classrooms who have the daily experience of interconnectivity and interactive media, which hooks them into the realms of e-mail, online games, networked information systems, and other small-scale, short-attention-span environments. Literary texts, particularly historical works and works of experimental innovation in the tradition of the avant-garde, appear to be peculiar artifacts—remote, antiquated, or esoteric. More crucially, dialogue and self-conscious reflection are barely present in the structure of media discourse with its emphasis on the consumable commodification of information and experience. Intellectual tools of critical reflection, so central to the mission of humanities education, seem increasingly important as survival skills. Within the academic community, an attitude of resistance to media (except as an object of study conceived within carefully controlled terms) prevents the serious integration of new technology into our activities. Particularly suspect are the forms that take advantage of the very features of mediated experience that are the daily currency of communicative exchange in the realms of entertainment, communication, business, and public and private news media. Ivanhoe and *SpecLab* projects are designed to take advantage of these forms by incorporating the modes of information access and use to which daily users of networked media are accustomed into academic study. Our work is premised on the idea that such a project is not a devil's bargain, but is in fact a significant gesture towards giving the humanities a viable profile for future generations.

Every generation asks certain questions of its literary and cultural heritage. In our own era, deconstruction has left a legacy of meta-questions that inform our research. Earlier twentieth-century literary critics addressed questions of style, authorship, attribution, meaning, and interpretive relations to ideology, politics, or culture. In the last quarter of a century we have added an attentive interest in asking questions about the ontology of texts, the intersubjective condition of their production and reception, and the ways their material existence is contingent upon a discourse field as an aspect of their capacity to function as elements within a signifying practice. The metaphors of networked culture find a corollary in the dispersed condition of discursive practice and in the contingent condition of texts within a diffuse field of artifacts. Nonetheless, students of literature and even many scholars of renown seem to forget these lessons when they sit down to the daily activity of interpretation. Students regularly come to the classroom intent on finding the "meaning" of a poem within an apparently stable text as if it were a self-evident and self-identical work.

Such attitudes prevail in the visual arts and media studies as well. A persistent strain of criticism regularly produces descriptively based analysis (hardly worthy of this final term) in the commercial press venues that serve the visual arts and narrative recounting based on story, plot, and character description in the popular press that follows the cinematic arts. The ideological and epistemological interlinkings of deconstruction have a hard time of it under such circumstances. While lip service to the theoretically informed agendas of critical inquiry persist in the research university, the goal of doing literary studies or pedagogical work that genuinely engages theoretical principles remains largely elusive. A more pressing concern, however, is the extent to which academic culture has become remote from popular and public experience. The humanities are equally threatened on several fronts. Popular disinterest in cultural traditions (beyond any but the most banal level of "produced" culture in an entertainment industry) is rampant. This disinterest is fiendishly matched by the esoteric self-involvement and oblivious condition of academic institutions (those at the highest level of research and seemingly the most immune to changes in the cultural context of their activity).[3]

Is it really possible that Ivanhoe can address this situation? Ivanhoe is specifically designed to take advantage of electronic instruments to teach literary (and other humanities) studies. Various activities contribute to Ivanhoe's overall agenda: bibliographical approaches to the material condition of textual production within a discourse field; the promotion of self-conscious reflection as a feature of critical interpretation; an engagement with the inter-subjective condition of scholarly and pedagogical work; the promotion of creative, individual identification with works of literary and cultural imagination; and serious engagement with ludic approaches to humanities scholarship.

Bibliographical studies have been much eclipsed in recent decades and spun off into a specialized study. The trends toward interpretation that form the core of much cultural studies work, to take one example, frequently promote "readings" of works (and particularly, their ideological subtexts) at the expense of attention to the material history of their production.[4] Generalization of the concept of "text," whereupon it applies to any instance, edition, form, or format of a particular work, abandons the critical ground on which bibliographical studies sought to establish parameters for literary study. More particularly, the approach to textual criticism that suggests that any "text" must be examined in the fuller discursive field of the artifacts comprising its production in the largest sense has been duly noted but put aside. This bibliographical approach requires a set of skills that critical "reading" does not always demand. The sheer difficulty of getting access to the scattered field of drafts, manuscripts, proofs, correspondence, and other documents that comprise a discourse field (or its partially delimited zone—since by definition such a field is infinite in its associational links) can be off-putting. A rarely asked question (to which many fear the answer) is how many scholarly projects take their form for reasons of convenience or incidental access rather than through considered decisions about the bibliographical foundation on which they are conceived. For the undergraduate imagination, the field of bibliographical studies has all the appeal of crumbling dust and fog machines. Nor do "bib studs" (bibliographical studies) have much glamour by contrast to such trendy approaches as "cult studs" (cultural studies) in graduate culture. Bibliography is regarded as a possible elective subfield, not as a necessary foundation for scholarly work of any and all kinds. Promoting the intellectual excitement of scholarly research—the mystery, the clue, the puzzle, and the pursuit of artifacts and documents—is a crucial aspect of Ivanhoe's design.[5]

Ivanhoe is not, however, some recast version of the board game *Clue* or a dinner mystery for the amateur scholar. A founding premise, that of the discourse field, provides a conceptual ground on which Ivanhoe works. This premise, articulated explicitly in McGann's *The Critique of Modern Textual Criticism*, is that any single instance of a work must be read against and into the full network of instantiations, notes, versions, editions, sketches, and other elements that comprise its bibliographical field (which should not be understood as closed or as a priori to a text).[6] Other responses and interpretations of a work—a film version, fictive biography, rework, or transformation—extend this discourse field. Demonstration of these principles is difficult in a print environment—almost impossible—and creation a digital collection of such material has been the motivating force for McGann's work on Dante Gabriel Rossetti. The archive of over eight thousand files in a robust database environment of searchable texts, transcriptions, facsimiles, images, and editions has provided—through its scope and its director's attitude—a case study in the theoretical issues emerging in digital humanities.[7] Extending Ivanhoe to engage with an electronic project like the Rossetti archive will present technical and theoretical challenges for the project because the scale and variety of materials and their relation to the game space will need to be considered, as well the conceptual difficulty of navigating a complex bibliographical field in an intelligent manner. Both are the real-world problems of Internet-based research. Format compatibility, data management, and information filtering from potentially vast resources will require significant shifts in scholarly modes and will create still-to-be-solved issues of access and delivery mechanisms. The scholar's role will shift in response. An informed perspective will function as guide and commentator, providing a critical reading to available materials rather than serving to present absent materials in a synthesis, summary, or representation constrained by limits of reproduction capabilities. Scholarly activity may change profoundly as a result, in ways as yet unanticipated.[8]

One such change is toward collaborative and intersubjective fields of exchange. The isolated activity of scholarship—the monk-like, cell-bound study—is significantly altered by the operation of that desktop space as a portal to connected, communicative exchange, one that links scholars to each other as well as to heretofore inaccessible materials that are linked, searchable, and comparable in ways previously rendered awkward to the point of near-impossibility by distance and distribution.[9] Non-self-identicality of texts, the theoretical engagement with a discourse field, attention to bibliographical artifactuality and documentary evidence, and the trail of works through their production histories and reception conditions as marked in responses and versions are all critical issues that have contributed to the parameters on which Ivanhoe has been conceived.

The structure of game play reinforces these precepts. As mentioned, Ivanhoe's role-playing structure allows for creative and imaginative writing as well as critical or scholarly production. Our discussion of the game has contained much debate on whether a "forking paths" or a "consensual text" model will prevail as moves are made. Such questions carry nontrivial theoretical and technical implications, especially as the chief goal of Ivanhoe is to increase self-conscious awareness, discourage solipsism, and create a relation between a literary work and the discourse field of which it is a part. Constraints that encourage interaction and foreground the intellectual or imaginative strategies of individual players are of major importance to achieving this goal.

Motivations for reading are underscored by the fact that every player has to intervene in the text. My own reading of the Scott novel, rather slowgoing and frankly unenthusiastic in the first instance, became immediately charged when I knew I faced the task of rewriting its Romantic ending. Reading with an eye toward making a clever move to delight or pique one's fellow players in the game focuses reading considerably. This response seems important as an aspect of Ivanhoe's effectiveness since scholarship and criticism, as well as authorship of creative and imaginative works, certainly calls out for interactive response. A community of readers/players is formed in Ivanhoe, and that social context also serves as an evaluative witness to the role-playing acuity of each player. Skills in bibliographical work, wit, aesthetics, or composition are rewarded in versions of the game in which points and scoring systems are put into play.[10] Roles, or "embodiment metaphors," may be specific, as in my playing Isabel Arundel, betrothed of Richard Burton, and rewriting young Catherine's character in *Wuthering Heights* in order to act out feminist fantasies of independent adventure while she pined for the wandering Burton. Or they may be vaguely defined and come into focus over the sequence of plays, as in *The*

Turn of the Screw game in which I played an OuLiPo-inspired (i.e., OuLiPo is a contraction of Ouvroir de Litterature Potential, the name for a group of French experimental writers who formed their Workshop for Potential Literature in the 1960s) graduate student assistant to the compilers of the concordance reworking the text at every occurrence of Flora's name to reconfigure the work as a feminist protest of James's conception of the girl's sexual imagination. In this latter case, I had a sense of the strategy I would use to generate moves but not of the specific outline of the persona through which the texts were to be enacted. More and less erudite engagements are possible, but every move must be accompanied by a journal entry justifying the intellectual basis of the contribution from the point of view of the assumed role. The point of such self-conscious masks is to debunk the myth of authorial neutrality. The fact of our authorial conceits and constructed subjectivities, though well accepted as a critical and theoretical legacy of thirty years of critical theory, tends, like deconstructive techniques of reading, to fall away in the practical activity of critical writing. Ivanhoe requires that such practices be maintained through deliberate acts that recognize the historicized and particular identity of any writing position.

A theoretically informed electronic environment is difficult to conceptualize in any cognitive gestalt. However, the technological capabilities of computational instruments are envisioned to serve some significant functions as the game evolves. First, the challenge is to use Ivanhoe to provide a collaborative, Web-based interface for pedagogical and scholarly work that integrates electronic collections and traditional, print-based collections in an environment of interactive exchange. The capacity to link artifacts, documentary evidence, and other materials of scholarship to an argument about or within a text will have a dramatically different effect than that of paraphrase and citation. The reading experience, which is much dependent in print formats on standardization, editing, selection, and re-representation in the form of notes and appendices, can become a suspended field of artifactual references linked to the work through a series of reasoned moves.

At one point, we envisioned an additional technological intervention that would have relied more heavily on computational processing techniques and visualization of game play, with the techniques for navigation and analysis intimately related in a tightly structured interface. Designing Ivanhoe posed various challenges. Metaphors and visualization had to be developed to communicate the critical precepts of Ivanhoe. What, for instance, is the shape of a discourse field, and how may the fundamentally non-self-identical character of a text be communicated? Such concepts are not easily visualized because conventions for their representation have never been established. The interface metaphors for the game toyed with allusions to books, desks, and library environments as well as those that imagine the electronic *n*-dimensional space as a multifaceted and accessible realm of discourse. Early schemes of display made use of changing scale, layered texts, documents, work spaces, and various navigational devices created with visualization techniques specific to digital media.[11] Here, too, the conceptual and theoretical issues intersected with technological considerations in ways that are far from trivial. Every consideration of textual relations and the potential manipulations made possible by electronic space carried with it a charge of semantic value, of information to be read into the field of meaning production and used to inflect, guide, and even produce the reading.

Now functional, Ivanhoe is a tool and a research project—a means to provoke investigation of many textual and interpretive issues. What is a text? In what artifacts and formats does its meaning reside? What modes of reading may digital instruments provide—and can they make any bridge at all between the traditions of humanities in academic culture and the rapidly changing landscape of mass media? Immodest as it may sound, Ivanhoe was created with the conviction that such a project has a role to play in preserving the humanities and providing reading practices that do engage new ways of conceiving of our work within a discourse field of visual, textual, and time-based media artifacts where literacy is understood as an informed and skilled capacity to perform with self-conscious criticality within a knowledge field. Ivanhoe attempts to foster such practices, making use of digital media in ways specifically designed to serve the goals of humanities pedagogy and research.

Deliberately provocative, our work is grounded in the conviction that aesthetics is crucial to the productive stimulation of imaginative work in the humanities, as much as the arts, and in pedagogical as well as scholarly circumstances.

Notes

1. See Fraser, "From Chaos to Conflict," and *Time, The Familiar Stranger*, for good starting points.
2. See McGann, *Radiant Textuality*, for a developed discussion of critical issues in their intersection with digital humanities.
3. No doubt colleagues will protest this last assertion, pointing to the volumes of scholarly writing that engage seriously with deconstruction and its methods, but anyone involved in observation of the daily practices of pedagogy knows all too well how persistently the "mining for meaning" approach to reading continues to hold sway in the classroom.
4. Contrast the membership in the Society for Textual Scholarship and the Modern Language Association for one test of the gnat and elephant proportion of interest in bibliographical study.
5. Bethany Nowviskie's "BiblioLudica" course on book history and bibliographical scholarship, funded by the Delmas Foundation, has been an experiment in introducing these approaches to undergraduates in the Media Studies program at the University of Virginia.
6. See McGann, *The Critique of Modern Textual Criticism* and *The Textual Condition*.
7. The distinction between a collection and a Web site is crucial; the former is organized with a searchable, flexible information structure and the latter is a set of hyperlinked documents. The distinction between a collection and an archive is more nuanced though contestably operative here.
8. See Hockey, *Electronic Texts*, and Robinson, "New Directions."
9. Medievalists have been particularly receptive to digital media access as the leaves of manuscripts are often scattered in so many different locations that bringing them together for contrast or collation is itself a logistical nightmare somewhat countered by the access electronic instruments permit.
10. Points and scoring are among the many features of the game that are toggled on and off for play. They seem largely unnecessary, though in *The Turn of the Screw*, since we were testing a game design with a particular game economy built into it, the points were linked to "inkwells" needed for making moves, thus putting certain constraints into play in the structure of the game.
11. Look at the Image Archive in the Design section of the Ivanhoe Game in *SpecLab*. http://www.speculativecomputing.org/ivanhoe/design.html and http://www.speculativecomputing.org/ivanhoe/notes.html.

Works Cited

Fraser, James T. "From Chaos to Conflict." *Time. Order. Chaos: The Study of Time IX*. Madison, Conn.: International Universities Press, 1998.
———. *Time, The Familiar Stranger*. Cambridge: Massachusetts University Press, 1987.
Hockey, Susan. *Electronic Texts in the Humanities*. Oxford: Oxford University Press, 2000.
McGann, Jerome. *The Critique of Modern Textual Criticism*. Charlottesville: University of Virginia Press, 1983.

———. *Radiant Textuality*. New York and Hampshire, UK: Palgrave, 2001.

———. *The Textual Condition*. Princeton, N.J.: Princeton University Press, 1991.

Robinson, Peter. "New Directions in Critical Editing." *Electronic Text: Investigations in Method and Theory.* Ed. Kathryn Sutherland. Oxford: Oxford University Press, 1997. 145–71.

SpecLab @ UVA. "The Speculative Computing Laboratory." 1 July 2004 http://www.speculativecomputing.org/.

11

Dehumanization, Rhetoric, and the Design of Wearable Augmented Reality Interfaces

Isabel Pedersen

Wearable computers ("wearables") are computers that people attach to their bodies in order to augment their personal experiences. People wear computers for many reasons. Mobility is the most obvious benefit. Scuba divers wear them in order to measure and record information while swimming underwater. Soldiers use them in urban war zones, where standing still is not an option. People also wear computers so that they can *exist* in new ways. Wearables work in conjunction with augmented reality (AR), which is technology that offers people a virtual reality *in addition to* the real-world reality that they currently experience (Azuma 356; Azuma et al. 34). Unlike virtual reality, AR does not forego the real; wearers simply augment the real with a virtual experience. Wearables are unique not only for their extraordinary capabilities but also because they undergo social scrutiny during their state of emergence into a new medium of communication. They are not yet a mass-market, small tech device (we cannot pick up a wearable at Walmart, for example); however, as they emerge, the discourse surrounding them simultaneously challenges them in terms of their treatment toward humans (present and future). The discourse of "wearable AR," attempts to liberate people from dehumanizing results so common in current technology. To an extent, wearable AR answers a call for action by N. Katherine Hayles: "Our challenge now, it seems to me, is to think carefully about how these technologies can be used to enhance human well-being and the fullness and richness of human-being-in-the-world, which can never be reduced merely to information processing or information machines" ("An Interview/Dialogue with Albert Borgmann and N. Katherine Hayles").

Wearable AR discourses already exhibit signs of thinking "carefully" about technology and assessing how we can "enhance human well-being." Steve Mann's ideal "Definition of 'Wearable Computer'" is a treatise on how we can achieve human-centricity for the medium. However, I claim in this chapter that there is a gap between ideal definitions and actual designs. Despite a surrounding discourse that strives for human-centricity, some wearable artifacts dehumanize people.

In a previous article, "A Semiotics of Human Actions for Wearable Augmented Reality Interfaces," I argue that the medium needs a design strategy that acknowledges the rhetorical nature of wearable interfaces. That article addresses the wearable's challenge to the human body (materiality), to the ways we make meaning, and to our ontological expectations. It uses Kenneth Burke's triad of terms for "order," drawn from *A Rhetoric of Motives*, as a theoretical framework. However, this chapter explores dehumanization and human-centricity to a much greater extent by addressing actual small tech devices.

The chapter has five parts. The first part explores the term "human-centricity" and compares it with Hayles's terms "posthuman" and "medial ecology" in order to reveal the high-level goals for the medium. The second part deals with Mann's definition because his terms inform the discourse of wearable AR; they offer a basis for ideal human-centric wearable AR. The third part presents the research problem, which is the gap between definition and actual design. This part presents the poma interface, which seems to emerge as a dehumanizing small tech artifact. The fourth part summarizes Burke's triad as a rhetorical design strategy in order to bring human-centricity back into the design cycle after we make definitions. And last, the fifth part presents an ideal wearable artifact, "CosTune," which fulfills human-centric goals by leveraging the rhetorical aspect of its interface. CosTune serves as an ideal design example.

What is Human-centricity?

Writers often judge the ideal qualities of computer interfaces through terms like "user-friendly," "intuitive," or "humane." "Human-centricity" is another trait that we commonly discuss and attempt to achieve in interface designs. Decisions that determine human-centricity occur early in the design cycle, long before we write software code or solder hardware components together. When designers simply ponder the uses, benefits, and risks of their inventions, they are determining how machinery will treat humans. In the most mundane, nontheoretical sense, wearables need to be human-centric because they involve individual, physical isolation. Wearing a wearable means strapping a device to a single human body. That wearable-toting body must be free to wander unhindered in the world. In this sense, the wearer needs to be an autonomous body in control of the computer. In a theoretical sense, I use the term "human-centricity" to mean a value system that privileges humans over machines and other hegemonic orders in order to avoid dehumanizing effects (e.g., machine-centricity, unwanted surveillance) (Pedersen, "A Semiotics of Human Actions for Wearable Augmented Reality Interfaces" 185; Pedersen, "Mobility, Human-centricity, and the Design of Wearable Augmented Reality Interfaces" 144–45). Most of the defining terms of wearable computers make this point overtly. They structure wearables to shift the balance of power to humans rather than machines.

In order to explore the term "human-centricity" further, I compare and contrast it with Hayles's notion of "posthumanism": "If my nightmare is a culture inhabited by posthumans who regard their bodies as fashion accessories rather than the ground of being, my dream is a version of the posthuman that embraces the possibilities of information technologies without being seduced by fantasies of unlimited power and disembodied immortality, that recognizes and celebrates finitude as a condition of the human being, and that understands human life is embedded in a material world of great complexity, one on which we depend for our continued survival" (*How We Became Posthuman* 5).

In this often-cited passage, Hayles strategizes future media designs. She posits two versions of the posthuman. *Nightmare* posthumans treat their bodies, their materiality, as superfluous, and they become "seduced" by fantasies of disembodiment through media; *dream* posthumans recognize their finitude and material condition as necessary, complex, and crucial for survival in conjunction with information technology. In her later work, *Writing Machines*, Hayles turns her attention to defining media as a process in the spirit of dream posthumanism. The term "medial ecology" suggests "that the relationships between different media are as diverse and complex as those between different organisms coexisting within the same ecotome, including mimicry, deception, cooperation, competition, parasitism, and hyperparasitism" (5). The ecological metaphor grounds the medial relationship in materiality.

In terms of the discourse of wearable AR, wearers are like dream posthumans because they acknowledge their material surroundings and their bodily movement in the world with utmost concern. Because a wearer exists outside of a virtual reality cave, he or she is subject to danger in the world like any other (nonvirtually enhanced) human being. The wearer depends upon a material existence for his or her "continued survival." However, dream posthumanism calls for humans to hinder their inclination to be "seduced by fantasies of unlimited power and disembodied immortality" ("How We Became Posthuman" 5) Wearable AR comes into conflict with dream posthumanism, in this case, because it *does* exhibit the desire to *better* humans with a virtual aspect. Neil Gershenfeld writes about this trait: "Wearable computers are a revolution that I'm certain will happen, because it already *is* happening. Three forces are driving this transition: people's desire to augment their innate capabilities, emerging technological insight into how to embed computing into clothing, and industrial demand to move information away from where the computers are and to where the people are" (47).

Gershenfeld captures the design intent behind the wearable. People keep striving to bring it into everyday existence because they want to augment innate capabilities and be *better* humans. The utopist intonation in this claim is unabashedly deliberate; wearable computing involves virtually augmenting one's subjectivity, one's reality, for what the wearer perceives as better. For Mann as a wearer, manipulating his reality with video *betters* his personal existence (Mann and Niedzviecki 3). The desire to augment forms a crucial aspect of human-centricity in the context of wearable AR. We need to address the design cycle in such a way that we not only avoid dehumanization as this medium emerges but also allow for reality-altering experiences that strive for utopist ends in appropriate ways.

This chapter is about changing a medium before it emerges. At times, the futurist vision of wearable AR smacks of so-called liberal humanist categorization because the subject seems to be able to transform significant regimes with autonomy (e.g., subject over object). However, this view misconstrues the situation. Currently, inventors like Mann understand desktop computing as antihuman or dehumanizing because hegemonic value systems like machine-centricity govern how computers interact with people (e.g., object over subject) ("Definition of 'Wearable Computer'"). He tries to shift the balance of power by offering new definitions of computing devices that are different from traditional computers. The discourse recognizes individuals as social agents with some capability to act or, as Anthony Giddens writes, with the capacity to "make a difference to a pre-existing state of affairs" (14). Ultimately, these interventionist definitions seek to structure a more cooperative relationship between media and social agents akin to a notion of medial ecology.

Definition and Discourse of Wearable AR

Few people have ever worn a computer or even seen a wearable. For most people, wearable AR is more a discourse than an obtainable object. Glenn Stillar comments on the way "discourse" acts upon people: "Discourse is 'action': It does things for social agents in the real context of their living. No discourse takes place outside the situated, embodied experiences and interests of the participants involved in an exchange. Discourse is an integral part of the complex goings-on that make up social life" (5).

Discourse occurs through social motives. It involves action and the degree to which participants enable or constrain action in the world through the creation and use of texts. The discourse of wearable AR causes and draws upon myriad texts that instantiate the discourse. These texts range from actual wearable inventions and the strategies behind them, to the laws that prevent them, to any mention of them in any media forum and countless other sources.

Currently, Mann is the most famous inventor of wearables and central to the discourse. Most introductory blurbs about wearables begin with a description of him and his WearComp (wearable computer) that he began devising in the 1970s. However, Mann also contributes a valuable and extensive terminology to the discourse that *leads* the designs of other inventors. He writes about computers and assistantship:

> Can you imagine hauling around a large, light-tight wooden trunk containing a co-worker or an assistant whom you take out only for occasional, brief interaction. For each session, you would have to open the box, wake up (boot) the assistant, and afterward seal him back in the box. Human dynamics aside, wouldn't that person seem like more of a burden than a help? ("Wearable Computing: A First Step toward Personal Imaging")

Out of his tongue-in-cheek analogy come two criteria for a wearable that appear repeatedly in his writing and the discourse in general: "constant" and "personal." "Constancy" means that wearables may have "sleep modes," but they are never *dead* like other small tech devices (e.g., laptops, cell phones, or personal digital assistants [PDAs]) that must be "opened up, switched on, and booted up before use" ("Wearable Computing as Means for Personal Empowerment"). A wearable is always *on*. The second criterion, "personal," breaks down to three supporting concepts. First, a wearable is a "prosthetic" and one should be able to configure it "to act as a true extension of the mind and body" ("Wearable Computing as Means for Personal Empowerment"). Second, a wearable is "assertive," which bars other people from asking a wearer to remove this device. People can be asked to remove a headphone or shut down a laptop, but a wearable needs to be exempt from this sort of request ("Wearable Computing as Means for Personal Empowerment"). Third, a wearable device is "private." No other person, server, or device can control the wearable unless the wearer turns over control. The privacy theme runs throughout all of Mann's writing and, to an extent, drives his desire to reinvent the way we compute ("Wearable Computing as Means for Personal Empowerment").

In addition to "constant" and "personal," Mann's "Definition of 'Wearable Computer'" offers further terms to define wearables:

1. Unmonopolizing
2. Unrestrictive
3. Observable
4. Controllable
5. Attentive
6. Communicative

In the following paragraphs, I discuss the design intent behind these terms because they reveal the human-centric nature of wearable AR. The *unmonopolizing* criterion eliminates most other small tech devices from Mann's wearable definition. A desktop computer, a laptop, a PDA, and even a wrist-worn computer monopolize the attention of the user. One has to focus eyes and ears upon them in order to use them properly. A virtual reality game is the most monopolizing because one physically situates the body in a fictive world, not only distracting one from the rest of reality but also restricting access to the real world. By insisting that the computer is unmonopolizing, Mann's wearable assumes mobility at the early design stage.

Working in concert with unmonopolizing, the *unrestrictive* criterion deals with the notion of physical constraint. Mann expects physical liberation for wearers. Not only does he mean that the wearable is wireless and not attached to a server or power source but he also intends that one can communicate while carrying out another physical activity (e.g., type while jogging) ("Definition of 'Wearable Computer'").

Observable means that one can always observe the computer. It lingers in the background, waiting to be called upon by the wearer. Like an aspect of the body, the wearable can always get one's attention if necessary. Some wearables are built into suits of clothing and facilitate activity in extreme environments like avalanche rescue missions.

Controllable ensures social agency over the computer. Humans ought to control computers and not the other way round. As a mild backlash against artificial intelligence (AI), Mann embraces "Humanistic Intelligence (HI)" (Mann and Niedzviecki 30), which stresses the human subject over the machine object.

Attentive and *communicative* soften some of the isolationism that the other criteria bring to the wearer. Attentiveness recognizes an environment to which the computer must *attend*; it suggests context. *Communicative* suggests a social network. The computer should allow "the wearer to be expressive through the medium, whether as a direct communications medium to others, or as means of assisting the production of expressive media" ("Definition of 'Wearable Computer'"). CosTune, discussed later in this chapter, exemplifies the idea of an "expressive" wearable.

Mann makes great strides toward humanizing wearable AR with his rich terminology. Through these terms, his wearable promotes the physical freedom to move unhindered. The body, no longer bound to the physical *withinness* of the lab, office, or virtual reality cave finds freedom in the *withoutness* of anywhere the wearer wants to go. In terms of physical movement, the wearable exudes centrifugality, an outward force radiating from the body. However, Mann's writing also implies centripetality, or an inward force, because it attempts as much as possible to isolate the wearer's body from other bodies and social connectivity. "Unmonopolizing," "unrestrictive," "observable," and "controllable" point to the inward force of the individual that does not necessarily want to partake in social systems. Despite the last criterion, "communicative," Mann's wearables structure a high level of isolation for the wearer. There exists no *wearable built for two* for Mann, likely because he insists that personal privacy is so important.

Despite this definition of wearables and a discourse that supports it, some inventors do not invent according to any human-centric definition. Conversely, some inventors claim human-centric ends and definitions but might not meet them with designs.

The Problem: Some Actual Wearable Devices Dehumanize

Apple Computer's iPod shuffle is not a wearable computer in an ideal sense. It is a personal digital music player that mimics not only the miniature proportions of the wearable but also some of the conceptual strategy. The iPod shuffle draws consumers with catchy taglines like "life is random" and "give chance a chance" (Apple Computer). It strategizes according to a nonscripted, human "life" rather than strategizing design according to machines as *orderers* of life. The iPod shuffle rhetoric demonstrates how discourses send out tendrils of human-centric influence within medial ecologies.

However, when one analyzes interfaces beyond advertising taglines, one finds that some of them are dehumanizing artifacts. Xybernaut's poma, "a mobile, Personal Multimedia Appliance" fits the general description of a wearable device. It is tetherless. It comes with a computer that one can attach to the belt. The mouse fits in one hand, and the thumb moves the cursor. It uses a see-through, head-mounted display over one eye. The poma operates on a customized desktop Windows interface, specifically, a version of Microsoft Windows' CE operating system (see Figure 1). The interface remediates the

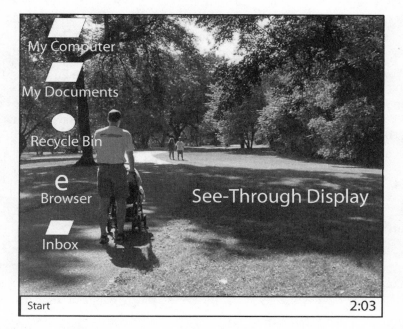

Figure 1. Mockup of poma's interface from the wearer's point of view

desktop interface for a nondesktop context. In doing so, the interface limits meaning-making rather than promotes it.

The desktop interface, a cluster of metaphors, interferes with wearable mobility. Many of the interface metaphors, which shape foreign experiences in familiar terms, limit understanding rather than facilitate it. The word *window*, for example, comes from the Norse *vindauga* and implies both wind (*vindr*), or the movement of wind from one place to another, and eye (*auga*), as if a window is really the anthropomorphized eye of a building staring outward onto another geography. "Window" suggests interaction between two worlds: an inside realm and an outside. In terms of an interface, "window" implies that I am on one side and the world of my computer with its files, folders, memory, and the Internet is on the other, which is a natural extension of what it feels like when one sits in front of a desktop computer. However, when one wears a display on the head, like glasses, the metaphor completely falls short. The ergonomic experience of a see-through display offers the chance to combine the real and virtual in a seamless manner. The windows metaphor prevents that seamlessness. Further, viewing the outside world through a desktop interface is akin to imagining that one sees the world through a small square. If one cuts a square border out of cardboard and holds it in front of the eyes, the view frames the world like a little picture, devoid of a periphery, devoid of places that exist at the back of the head. Ultimately, one cannot inhabit one's desktop interface.

The metaphoric world of the desktop is two-dimensional, unlike the three-dimensional world of humans. It suggests a space on which one places files and folders; it is a flat world inspired largely by the experience of paper. It consists of buttons that imply depth and persuade the user to act, but the buttons do not provide enough depth to inspire the idea of a realm. They are bumps on a flat landscape. Even metaphorically, one cannot go and sit upon the desktop or beside it or even under it. If one wears the poma to view a flock of migrating swans, one must deal with the metaphoric jarring of office files, folders, trashcans, windows, and textual labels amid the organic world of the swans. A mobile "desktop" is a difficult metaphoric blend. No one slings his desk on his back and goes for a walk. Overall, the poma's remediation of the desktop interface cripples the wearable experience in terms of human meaning-making.

Despite all the hardware and ergonomic features of the poma, it neglects to strategize a new interface. In a sense, it acquiesces to a remediation. More importantly, evidence of Mann's ideal terms do not seem to inform the interface design; in fact, the poma promotes the opposite situation. While one might be able to walk freely, the desktop interface is monopolizing, restrictive, distracting, *un*controllable, *un*attentive, and only marginally facilitates communication with others through its Web features. We need to hold wearable AR to its promises of human-centricity and provide a means to that end.

A Rhetorical Means to Bring About Human-Centricity

Burke writes that rhetoric offers the opportunity of "inducing cooperation in beings" through language or symbol systems (43). "Inducing cooperation" means transformation rather than persuasion. Stillar writes that "rhetoric deals with language's role in identification and division among social agents; it focuses on the exchange of discourse as a central mode through which social orders are constructed and transformed through addressed symbolic action" (62). Humans exist in a state of division—we forever agree and disagree. However, through language (i.e., symbol systems) we adhere to constructed social orders and partake in transformation. When the Microsoft Web site slogan asks "Where do you want to go today?" it induces cooperation with all its Web visitors hailed by the question. It structures a social order in the interface that implies that Microsoft could *take* anyone anywhere. It structures desire in the visitor *to go somewhere*. While Microsoft uses interface rhetoric for commercial purposes to lead its visitors, the possibility exists to leverage rhetoric for human-centric ends if we strategize accordingly.

In "A Semiotics of Human Actions for Wearable Augmented Reality Interfaces," I use Burke's terms for order, a triad—the "positive," "dialectical," and "ultimate" (183–89)—to serve as a framework for design, and I summarize that framework here. Burke's first term of order, the positive, relates to a physicalist vocabulary: "First, we take it, there are the *positive* terms. They name par excellence the things of experience, the *hic et nunc*, and they are defined *per genus et differentium*, as with the vocabulary of biological classification. . . . In Kant's alignment, the thing named by a positive term would be a manifold of sensations unified by a concept" (183).

The positive refers to things in utter *disambiguity*. Positive terms name things with a sensory existence like rocks, books, tables, tastes, and sounds. Wearers signify according to positive vocabularies because they move around the real world and deal with things of a tangible existence. When the wearable reads temperature, latitude, or speed, for example, it deals with a positive terminology. Mann's WearComp lets him see with stroboscopic vision to freeze motion and, as he says, "count the grooves in the tread" of a wheel spinning at a hundred kilometers an hour (Mann and Niedzviecki 3). Stroboscopic readings are positive terms.

The second order of the triad is the dialectical: "Here are words that belong, not in the order of *motion and perception*, but rather in the order of *action and idea*. Here are words for *principles* and *essence*. . . . Here are titular words. Titles like "Elizabethanism" or "capitalism" can have no positive referent, for instance. . . . You define them by asking how they *behave*; and part of an expression's behavior . . . will be revealed by the discovery of the secret modifiers implicit in the expression itself" (Burke 184–85). The dialectical refers to debatable concepts or ideas that deal with an either/or choice. Not solely determined by systems, social agents have the ability to make choices within contexts. Computer interactivity is very much of the dialectical because it is about constant choice-making. "Game-City," for example, is a multiuser wearable game application (Cheok et al.). Players wearing head displays traverse areas as big as cities interacting with other people, fighting virtual witches, and gathering virtual treasure. The game involves unknown landscapes and constant decision making between people (e.g., choices). Interactivity is a dialectical phenomenon because it deals with the way participants exchange meaning among each other and with themselves.

The third order, the ultimate, is instantiated by terms that resolve the conflict of the dialectical order:

> Now, the difference between a merely "dialectical" confronting of parliamentary conflict and an "ultimate" treatment of it would reside in this: The "dialectical" order would leave the competing voices in a jangling relation with one another . . . but the "ultimate" order would place these

> competing voices themselves in a *hierarchy*, or a *sequence*, or *evaluative series*, so that in some way, we went by a fixed and reasoned progression from one of these to another, the member of the entire group being arranged *developmentally* with relation to one another. . . . The "ultimate" order of terms would thus differ essentially from the "dialectical" . . . in that there would be a "guiding idea" or "unitary principle" behind the diversity of voices. (Burke 187)

Ultimate terms structure as a guiding principle, a seeming authority, or a hierarchy. Any interface design, for example, is also a conventional agreement by its participants to think and act in a certain manner. As discussed earlier, Microsoft's hailing "Where do you want to go today?" structures Microsoft as a guiding or "ultimate" principle, and all its Web visitors are submissive subjects. When dealing with a reality-based medium, ultimate terms also structure ontological expectations, or "beingness," the ways we construct our own reality. If wearers can perform tasks that are extraordinary, like seeing a microscopic bug ten feet away, they not only have an augmented sense but they also *exist* in a new way by signifying with their interfaces.

The triad of terms enables us to realize the transformational process that occurs in texts: "In an ultimate dialectic, the terms so lead into one another that the completion of each order leads to the next . . . there must be a principle of principles involved in such a design—and the step from the principles to a principle of principles is likewise both the fulfillment of the previous order and the transcending of it" (Burke 189). Each term encompasses the next and transcends it. Positive vocabularies transform to dialectical vocabularies; ultimate vocabularies *reorder* both the positive and the dialectical. The wearer moves in the real world (positive), causing interaction with *others* (dialectical) according to a guiding idea like a desktop interface that governs the other terms (ultimate). I argue that *beingness* offers the most desirable hierarchy for the wearer. Instead of strategizing the interface according to hegemony (an external *orderer*), we ought to strategize according to human-centricity as a principle. We need to manipulate the positive, dialectical, and ultimate potentials of the interface so that humans govern the wearable experience themselves according to their own senses of being.

CosTune

"CosTune" serves as a model to which wearable AR can and should aspire in terms of rhetorical transformation (see Figure 2). Invented by Kazushi Nishimoto and others, CosTune is a wearable musical instrument (Nishimoto et al.). The interface is a piece of clothing rigged with sensors that lets roving wearers listen to music, create it, and share it with other wearers through regional servers. However, these designers do more than set down a blueprint for hardware and software. CosTune demonstrates a strategy that (1) supports a physical challenge to the body in *positive* terms, (2) alters the way humans make interactive meaning with others and as solitary beings in *dialectical* terms, and (3) challenges ontological expectations of existence. *Beingness* serves as the principle of principles; it functions in *ultimate* terms.

One journalist ruminates on the role CosTune serves:

Figure 2.   Kazushi Nishimoto wearing CosTune (Courtesy of Kazushi Nishimoto.)

Riding the subway, walking through the mall, waiting at the airport lounge, a tune plays inside your head. You want to hum, hum out loud, really, to share it with someone.

Of course, you can't. It's not very polite. Especially when the man next to you is deeply engrossed in a crossword puzzle. Who cares about your little private symphony anyway? (Bhattacharjee)

CosTune allows one to be silent around an inadvertent neighbor while at the same time listening and creating music with others. It can act like a regular instrument whereby a tap on jacket, pants, or gloves relates to a note of music or it can project phrases of music. A phrase is a "song component of a certain length, and can be divided in time and/or instrumental type or role. Conversely, a song is defined as an organized set of phrases" (Nishimoto et al. 56–57). In other words, one need not be a trained musician to use CosTune. Humans use memory to remember music, and transformation of this memory to phrases of music is the only requirement for CosTune. In addition, CosTune encourages nonmusicians in other ways. One uses the hands (e.g., clapping), the feet (e.g., tap dancing), or the voice (e.g., singing) to make musical sounds. CosTune mimics this condition because one taps out the phrases on the tactile sensors. Instead of alienating humans with a foreign input device, CosTune embraces a common human process.

CosTune encourages not only mobile listening but also mobile, collaborative participation in music-making: "[CosTune is a] wearable system of sensors that allows users to create and play music while doing routine tasks and to share performances with one another over a wireless network. The performers hear the music in their headsets . . . the keyboard is reduced to a light interface, like a pattern of metallic thread woven into fabric" (Bhattacharjee). CosTune is more than a wearable musical instrument because it achieves a roving interconnection of music between wearers: "We designed 'CosTune' to be a communications tool rather than a simple musical instrument. The most significant feature of CosTunes is that they are equipped with wireless network functions. This allows users to communicate with anyone, anytime and anywhere by means of music in an ad hoc manner without scattering sound. We think that this feature makes CosTune a supporting tool for forming communities in the real world" (Tada et al.). Rather than claiming that CosTune promotes collaboration, the designers step to the next level, the purpose for collaboration, which is *using music to form community*. The music and the wearable device enable the social experience. CosTune is respectful of a "pluriverse." Running counter to some of the ideals for a wearable, CosTune relies on a remote server so that it can pass musical information from one wearer to the next in local regions. Wearables imply the *untethered* experience, but CosTune is somewhat tethered to these local servers. However, the placement of servers that link wearers together (because they must be in the range of this server) coincides with a view of CosTune as a regional application.

The designers list many possible real-life applications for CosTune, but one of them is strikingly unique:

> [People] generate the "atmosphere of the region" and the atmosphere of the region attracts those who like the atmosphere. As a result, regions acquire unique characteristics, e.g., SoHo in N.Y. and Harajuku in Tokyo. We think that the music that is performed in a region must reflect the characteristics of that region. Conversely, the jam sessions performances and the composed musical pieces must become different depending on the regions where a CosTune user visits. Therefore, we think people who want to enjoy the music of a specific region should actually (not virtually) visit the region and meet the people of the region. (Nishimoto et al. 61)

Music contributes to the fabric of a local community. CosTune augments regional rather than global culture. Nishimoto's strategy indicates a trend in wearable discourse that privileges the geographically close over the virtually *anywhere*. More specifically, this movement does not encourage virtually linking people from everywhere as so many other virtual projects do, like the Internet. CosTune augments the wearer in terms of local culture.

One wonders, however, about stagnation. Will the musical experience and the resulting community become stagnant if the CosTune wearers only experience each other's music locally with no outside influence? Rather than hook all the servers together in one massive network, the design motivation behind CosTune is to let wearers pass phrases of music to other regional servers when they wander near them in a "phrase scattering mode" (Nishimoto et al. 61). The tethering nature of CosTune does not cause a closed context because the technology mimics human social interaction in a more organic manner: "phrases can be transmitted by CosTune users moving around. We think it is rather interesting to isolate the servers than to directly connect them via a network. As a result, phrase transportation among servers can be achieved only when CosTune users move. In other words, the users become 'vectors' of regional music, like butterflies carrying pollen among flowers. Depending on the human flow, the regional musical cultures can be hybridized" (Nishimoto et al.).

CosTune music travels in the real world via the wandering human, mimicking the real transfer of messages between people. These designers cast the metaphor of plant reproduction overtop this communication exchange to emphasize the way it aspires to the organic, the real, and not the omnipresent possibility of the virtual, which can send files everywhere at anytime.

Mobility becomes the catalyst for hybridity or change in this system; it brings newness. As Gregory Bateson says, "in contrast with epigenesis and tautology, which constitute the worlds of replication, there is a whole realm of creativity, art, learning, and evolution, in which the ongoing processes of change *feed on the random*" (49; Bateson's italics). The random movement of CosTuners brings about a newness similar to a stochastic system whereby "a sequence of events combines a random component with a selective process," limiting the randomness but allowing enough to bring change to the system (Bateson 249). Interestingly, mobility is synonymous with interactivity in this configuration (i.e., a person places herself in the proximity of a new wearer or server to pass phrases); consequently, movement causes transformation within the sign system.

In terms of Burke's triad, this movement (positive order) causes interaction with other music-makers (dialectical order). However, chaos does not ensue, and those phrases do not jangle about without resolution. The meeting of musical phrases brings a hybrid, an offspring, or a blend, constituting a resolution (ultimate order). The new musical texts bring about unity and a new design. CosTune wearers live according to an ultimate order that facilitates a new beingness. The roving, composing, one-wearer-band/virtual symphony musician uses the modality of music to *be* in a new way. Overall, CosTune does attempt to close the gap between definition and design by meeting ideal criteria like unmonopolizing, unrestrictive, observable, controllable, attentive, and communicative. It attempts to "induce cooperation" among social agents and, in doing so, achieves a much more ecological experience for the wearer.

Much of this chapter has been about how we can influence an emerging medium to better people's lives instead of diminishing them. To an extent, the medial relationship in which wearable AR and the desktop interface coexist has caused dehumanizing results. This chapter argues that we need to design human-centric interfaces in the future, and

CosTune is one ideal example. As wearables become smaller and smaller, integrating more with our bodies and brains, we will need a conceptual design strategy that does not let our machines wear the pants.

Works Cited

Apple Computer. "iPod shuffle." *Apple.* 2005. 9 May 2005 http://www.apple.com/ipodshuffle/.

Azuma, Ronald. "A Survey of Augmented Reality." *Presence, Special Issue on Augmented Reality* 6.4 (1997): 355–85.

Azuma, Ronald, et al. "Recent Advances in Augmented Reality." *IEEE Computer Graphics & Applications* 21.6 (2001): 34–47.

Bateson, Gregory. *Mind and Nature: A Necessary Unity.* New York: Bantam, 1979.

Bhattacharjee, Yudhijit. "Making the Music Sway to Your Beat." *New York Times* 29 Nov. 2001, sec. Technology. 6 February 2006 http://www.rpi.edu/web/News/NYTSwaynp.html.

Burke, Kenneth. *A Rhetoric of Motives.* 1950. Berkeley: University of California Press, 1969.

Cheok, Adrian David, et al. "Game-City: A Ubiquitous Large Area Multi-Interface Mixed Reality Game Space for Wearable Computers." *Sixth International Symposium on Wearable Computers.* Seattle: IEEE, 2002. 156–57.

"CosTune Jacket Type." *Nishimoto Laboratory.* Japan Advanced Institute of Science and Technology. 14 Dec. 2002 http://www.jaist.ac.jp/ks/labs/knishi/CosTune.html.

Gershenfeld, Neil. *When Things Start to Think.* New York: Henry Holt, 1999.

Giddens, Anthony. *The Constitution of Society: Outline of the Theory of Structuration.* Berkeley and Los Angeles: University of California Press, 1984.

Hayles, N. Katherine. *How We Became Posthuman: Virtual Bodies in Cybernetics, Literature, and Informatics.* Chicago: University of Chicago Press, 1999.

———. *Writing Machines.* Cambridge, Mass.: MIT Press, 2002.

"An Interview/Dialogue with Albert Borgmann and N. Katherine Hayles." *University of Chicago Press.* 1999. 10 May 2004 http://www.press.uchicago.edu/Misc/Chicago/borghayl.html.

Mann, Steve. "Definition of 'Wearable Computer.'" *Wearcam.org* 12 May 1998. 17 June 2003 http://wearcam .org/wearcompdef.html.

———. "Wearable Computing: A First Step toward Personal Imaging." *Computer* 30.2 (1997): 25–32.

———. "Wearable Computing as Means for Personal Empowerment." *Keynote Address presented at the 1998 International Conference on Wearable Computing.* Fairfax, Va.: ICWC, 1998. 17 June 2003 http://wearcam .org/icwckeynote.html.

Mann, Steve, and Hall Niedzviecki. *Cyborg: Digital Destiny and Human Possibility in the Age of the Wearable Computer.* Toronto: Doubleday, 2001.

Nishimoto, Kazushi, et al. "Networked Wearable Musical Instruments Will Bring a New Musical Culture." *The Fifth International Symposium on Wearable Computers.* Zurich: IEEE Computer Society, 2001. 55–62.

Pedersen, Isabel. "A Semiotics of Human Actions for Wearable Augmented Reality Interfaces." *Semiotica* 155.1–4 (2005): 183–200.

———. "Mobility, Human-centricity, and the Design of Wearable Augmented Reality Interfaces." *International Journal of the Humanities* 3.1 (2006): 143–54.

Stillar, Glenn. *Analyzing Everyday Texts: Discourse, Rhetoric and Social Perspectives.* Thousand Oaks, Calif.: Sage, 1998.

Tada, Yukio, et al. "Towards Forming Communities Using Wearable Computers." *International Workshop on Smart Appliances and Wearable Computing.* Scottsdale, Ariz.: IWSAWC, 2001. 260–65.

Xybernaut. *Poma User's Guide.* Fairfax, Va.: Xybernaut Corporation, 2002.

12

Sousveillance: Wearable and Digital Tools in Surveilled Environments

Jason Nolan, Steve Mann, and Barry Wellman

For decades, the notion of a "surveillance society" where every facet of our private life is monitored and recorded has sounded abstract, paranoid or far-fetched to some people. No more! . . . Yet too many people still do not understand the danger, do not grasp just how radical an increase in surveillance by both the government and the private sector is becoming possible . . . from a number of parallel developments in the worlds of technology, law and politics.

—Jay Stanley and Barry Steinhardt,
"Bigger Monster, Weaker Chains"

Sousveillance: Surveilling the Surveillers

Surveillance is everywhere, but often little observed. Organizations have tried to make technology mundane and invisible through its disappearance into the fabric of buildings, objects, and bodies. The creation of pervasive ubiquitous technologies—such as smart floors, toilets, elevators, highway cameras, and light switches—means that intelligence-gathering devices for ubiquitous surveillance are also becoming invisible (Mann and Niedzviecki; Marx "The Engineering of Social Control"; Lefebvre). For example, closed-circuit television networks (CCTV) surveill neighborhoods in the name of public safety. This proliferation of *small* technologies and data conduits has brought new opportunities for observation and data collection, making public surveillance of private space increasingly ubiquitous. All such activity has been *sur*veillance: organizations observing people.

One way to challenge and problematize both surveillance and our acquiescence to it is to resituate these technologies of control on individuals, offering panoptic technologies to help them observe those in authority. We call this inverse panopticon "*sous*veillance," from the French words *sous* (below) and *veiller* (to watch). With many people in the developed world carrying camera-equipped mobile phones, ordinary people now have ample means for portable, low-cost, easy to use sousveillance. Web sites such as YouTube (www.youtube.com) for videos and Flickr (www.flickr.com) for photographs, make it almost as easy for people to share sousveillance with others, one-to-one, as it is to broadcast to the public. Sousveillance can be a form of tactical media activism. As such, it can be seen as a proven mode of resistance. For example, police *agent provocateurs* were quickly revealed on YouTube when they infiltrated a demonstration in Montebello, Quebec against the leaders of Canada, Mexico, and the United States (August 2007). When the head of the Quebec police publicly stated that there was no police presence, a sousveillance video showed him to be wrong. When he revised his statement to say that the police *provocateurs* were peaceful observers, the same video showed them to be masked, wearing police boots, and holding a rock (http://www.youtube.com/watch?v=St1-WTc1kow).

There are many similar examples, such as the widely-viewed YouTube showing of UCLA police attacking a student in the university library (Fall 2006: http://www.youtube.com/watch?v=AyvrqcxNIFs). Other projects include the "Yes Men" (http://theyesmen.org), RTMARC (http://rtmark.com), Next5Minutes.org (http://www.next5minutes.org); in the Critical Art Ensemble's court battles; and on various Nettime.org lists (Lovink and Schneider).

The sousveillance performances presented in this chapter are very much forms of activism. However, instead of mobile camera phones, the performances use specially created tools of inquiry specific to the form of intervention anticipated. As part of this tradition of challenging institutional control over public life, sousveillance is a form of "reflectionism" (Mann, "'Reflectionism' and 'Diffusionism'"), a philosophy and a procedure of using technology to mirror and confront social ecologies dominated by bureaucratic organizations by holding up a mirror and asking: "Do you like what you see?" It represents a methodology for exploring surveillance—in all its forms—and the dynamic relation between technology and its cultural ecology. Accordingly, reflectionism is a technique for inquiry-in-performance that is directed toward uncovering the panopticon, undercutting its primacy and privilege, and relocating the relationship of the surveillance society within a more traditional "commons" notion of observability (Ostrom).

Reflectionism is especially related to "détournement": the tactic of appropriating tools of social controllers and resituating these tools in a disorienting manner (Rogers; Ward). It extends the concept of détournement by using the tools against the organization, holding a mirror up to the establishment, and creating a symmetrical self-bureaucratization of the wearer (Mann, "'Reflectionism' and 'Diffusionism'"). In this manner, reflectionism is related to the "Theater of the Absurd" (Bair), and the Situationist movement in art. Reflectionism becomes sousveillance when it is applied to individuals wearing digital tools to observe the organizational observer. Sousveillance focuses on enhancing the ability of people to access and collect data about their surveillance and to neutralize surveillance, creating a new ecological balance. As a method to map and protect personal space, it resonates with Gary Marx's proposal to resist surveillance through noncompliance and interference "moves" that block, distort, mask, refuse, and counter-surveil the collection of information ("A Tack in the Shoe").

Reflectionism differs from those solutions that seek to regulate surveillance in order to protect privacy (Rhodes et al.). It contends that regulation of surveillance is as much a pacifier as a solution because in a regulatory regime, surveillance information is largely exchanged and controlled by external agents over which individuals have little power. For example, a recent regulatory proposal from the American Civil Liberties Union suggests "surveillance cameras . . . must be subject to force-of-law rules covering important details like when they will be used, how long images will be stored, and when and with whom they will be shared" (Stanley and Steinhardt 2). By contrast, reflectionism seeks to nurture a new relationship between the surveiller and the person being surveilled (i.e., the surveillee), including enabling the surveillee to surveil the surveiller.

Probably the best-known recent example of sousveillance is when Los Angeles resident George Holliday videotaped police officers beating Rodney King after he had been stopped for a traffic violation. The ensuing uproar led to the trial of the officers and serious discussion of curtailing police brutality (Cannon). Taping and broadcasting the police assault on Rodney King was serendipitous and fortuitous sousveillance. Yet planned acts of sousveillance can occur, although they are rarer than organizational surveillance. Examples include customers photographing shopkeepers; taxi passengers photographing cab drivers; citizens photographing police officers who come to their doors; civilians photographing government officials; and individuals beaming satellite shots of occupying troops onto the Internet. In many cases, these acts of sousveillance violate prohibitions stating that ordinary people should not use recording devices to record official acts. For example, many countries, including the United States, have prohibitions against photographing military bases. More often these prohibitions are unstated. For example, although many large stores do not want photographs taken on their premises, it is rare to see a sign prohibiting such photography.

The Rise of Neopanopticons

Privacy is a psychological as well as a social and political requirement. People seek control over the degree of anonymity they possess in their relationships by choosing what personal information to reveal to another person based upon their relationship (Ingram). Yet the asymmetrical nature of surveillance is characteristic of the unbalanced power relationship of a destabilized social ecology. The power that the police or customs officers assert when they search a person's belongings or the contents of his or her pockets—when the officers themselves cannot be searched—reflects a relationship firmly located in the panopticon that is seen in the asymmetric photography and video policies of the examined establishments themselves (Ingram).

However, the notion of ubiquitous surveillance is a longstanding outcome of institutionally organized social relationships. Jeremy Bentham's (1838) Panopticon defined a system of observation in which people could be placed under the *possibility* of surveillance without knowing whether they were actually being watched. Bentham proposed such an architecture for use in prisons, schools, hospitals, and workplaces. Michel Foucault describes the implications of this system:

[T]he major effect of the Panopticon [is] to induce in the inmate a state of conscious and permanent visibility that assures the automatic functioning of power. So to arrange things that the
surveillance is permanent in its effects, even if it is discontinuous in its action; that the perfection of power should tend to render its actual exercise unnecessary; that this architectural apparatus should be a machine for creating and sustaining a power relation independent of the
person who exercises it; in short, that the inmates should be caught up in a power situation of
which they are themselves the bearers. (*Discipline and Punish* 201)

Bentham's ideas represented an updating of governance techniques in preindustrial
societies for industrialized societies. The densely knit connections and tight boundaries of
preindustrial "door-to-door" communities fostered direct visual observation as a means
of social control (Ostrom; Wellman, "The Network Community"; Wellman, "Physical
Place and Cyberspace"). With the Industrial Revolution, societal scale increased beyond
the ability for little groups of neighbors to eye one another. There was a perceived need
for industrialized social control. Hence, panopticons employed hierarchies of organizational employees to observe public spaces in prisons, factories, and so forth. Indeed,
"panopticism was a technological invention in the order of power, comparable with the
steam engine" (Foucault, "Question of Geography" 71; see also Foucault, *Discipline and
Punish* and "Prison Talk").

In post-industrial societies, new communication techniques are exploited by neopanopticons. In public or semipublic (e.g., commercial) locations, individuals are liable to
become unwilling and sometimes unknowing subjects of surveillance, and the knowledge
that they may be under surveillance may be sufficient to induce obedience to authority
(Foucault, *Discipline and Punish*). Recall the television show *The Prisoner*, set in an ostensibly bright and cheery village under constant video surveillance. The surface pastoralism
of the village is enforced by a panopticon.

Since then, surveillance techniques have increasingly become embedded in technology. And as tools that can be used for surveillance, or to challenge surveillance, have
become embedded in personal digital assistants (PDAs) and Web-enabled camera phones,
the potential is at hand for ubiquitous sousveillance as an organized and systemic challenge to institutional surveillance. Where people once watched people with their naked
eyes, computer-aided machines now offer remote sensing of behavior. Automatic messages inform callers to organizational "call centers" that their conversations are being monitored to "improve customer service." Surveillance cameras are spreading rapidly through
public space; "[a] survey of surveillance cameras in Manhattan, for example, found that it
is impossible to walk around the city without being recorded nearly every step of the way"
(Stanley and Steinhardt 2). Video cameras can be almost invisibly small, communication
networks can direct surveillance images to monitors (both people and screens) located
elsewhere, and information technology can use facial recognition software to identify
likely suspects. Those subject to neopanopticons do not have direct visual and aural contact with those who are observing them.

People are subjects being monitored in two senses of the word. First, they are subjects
of observation on video monitors displaying and previewing the acquisition of their image.
In this sense, they are subjects of the camera (as in the "subject matter" of a photograph).
Second, subjects are under the potential control of people in positions of authority who
are organizational monitors of their behavior. They are like the subjects of a dictator, an

authority figure, or an organizational institution. And the situation created *by* technology-mediated surveillance creates an imbalance in terms of who can undertake these forms of surveillance.

There is a digital divide in the unequal access to these technologies by the general public. The proliferation of environmental intelligence, in the form of cameras and microphones observing public spaces, challenges the traditional ability of an individual to be able to identify and watch the watchers. The collection of data in public places, with the camera as the dominant form of data input device, is coupled with the integration of surveillance with statistical monitoring and security applications. The passive gathering of intelligence represents a challenge to privacy in public places that has been largely accepted (Mann and Guerra; Webster and Hood).

Even before the personal computer revolution, other efforts were made to attach computing technology to the body (Mann, "Wearable Computing"; Mann and Niedzviecki). Wearable systems have been developed to provide multimodal sensory interfaces as well as electrical connections to and from the body. These allow remote control of the body by wireless communications. For example, EyeTap technology has evolved as a system for causing the eye itself to function as if it were both a camera and a display (Mann and Niedzviecki). This has allowed the devices to modify visual perception, thus setting the stage for an interface to the body that challenges notions of free agency and locality of reference.

The Tools: Conspicuously Concealed, Wearable Flat-Panels, and the Invisibility Suit

More active forms of sousveillance confront surveillance by using wearable computing to surveil the surveillers reflectively, bringing into question the very act of surveillance itself. Because of the mobility of the modern individual, this act is best accomplished by mobile, wearable computers. In the mobile society of the early twenty-first century, Western societies move among milieus. Individuals' personal environments travel with them in the unstable environment of ostensibly neutral public spaces such as streets, sidewalks, and shopping malls (Lefebvre; Wellman, "Physical Place and Cyberspace"). In such milieus, individuals are largely responsible for their own security and integrity. Wearable computing devices afford possibilities for mobile individuals to take their own sousveillance with them. Given this frequent sociophysical mobility, it makes sense to invent forms of wearable computing to situate research devices on the bodies of the surveilled (e.g., customers, taxicab passengers, citizens). The act of holding a mirror up to the social environment, allows for a transformation of surveillance techniques into sousveillance techniques that watch the watchers.

Some of the performance devices described in this chapter also incorporate various large flat-panel display screens, worn on the body, that display live video from a concealed camera or a video recording from a previous trip to the same shop (see Figure 1). The ambiguity surrounding when the video was recorded allows the wearer to explore issues of recording and displaying video images in locations where cameras are prohibited.

When the performers wear a flat-panel screen or a data projector on their bodies, they show images of themselves to the people in the immediate environment. These visible displays can evoke social control without any need for comments from the wearer of

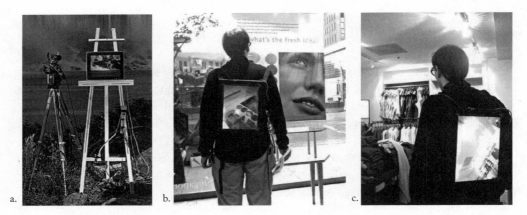

a.   b.   c.

Figure 1a.   A television mounted to an easel and placed at the base of Niagara Falls illustrates the principle of illusory transparency with a video camera connected to the television and positioned such as to show the subject matter behind the television in exact image registration with what one would see if the television were not present.

Figure 1b.   Invisibility suit: Street Theater of the Absurd. A display is incorporated into a wearable computer system. The back-worn display shows output from a front-worn camera, so that people can "see through" the wearer; the wearer's back is a window, showing what is in front of the wearer. When asked by agents of surveillance about this special clothing, the wearer responds that the device is an "invisibility suit to provide privacy and protection from surveillance cameras." This answer externalizes the locus of control: first, a silly idea such as an invisibility suit makes it hard for the agent to reason with the wearer; second, the wearer argues that the motivation for wearing the camera is to provide protection from being seen by surveillance cameras. The surveillance agent's objection to the sousveillance camera becomes an objection to his own surveillance camera.

Figure 1c.   Sousveillance under surveillance: the invisibility suit worn under a department store's ceiling domes

the camera and display. When the sousveiller remains mute until addressed by store employees, the wearable sousveillance devices become the object of attention rather than the person wearing them. Indeed, the probability of interaction increases with an increase in the overtness of the sousveillance camera or data collection device (Mann and Niedzviecki).

Many of the questions asked by viewers concerned visibility and transparency. In one performance, a display worn on the performer's back was arranged to show the view from a camera worn on the front. When asked what this device was, the performer simply said that it was an "invisibility suit" (see Figure 1). Obviously, this notion is nonsensical in the sense that the device certainly does not give invisibility. In fact, it attracted more attention. Presenting the camera as a form of theatre helped to legitimize it as an externality, although with less success than presenting the camera as an external corporate requirement. In some performances, when staff objected to the video displays, the performers offered to cover the displays with sheets of paper so that the images would no longer be visible.

Such situations create a distinction between the conflated issues of *privacy*—no personal data being collected—violated by input devices such as cameras versus *solitude*—no intrusion on personal space—violated by output devices such as video displays. Where privacy is the right to be free of the effects of measuring instruments such as cameras, solitude is the right to be free of the effects of output devices such as video displays.

Wearable Computing for Sousveillance

Digital technology can build on personal computing to make individuals feel more self-empowered at home, in the community, at school, and at work. Mobile, personal, and wearable computing devices allow people to take the personal computing revolution with them. These tools include camera phones, PDAs, arc lamps, wearable computers, hidden and visible cameras (still and video), night vision cameras, laptops, portable surveillance domes, wireless LANs (local area networks) that broadcast between sousveilling participants, broadcast technologies such as blogs, YouTube, and Flickr, and even technologies such as Google to search for and access unsecured surveillance cameras that are connected to the Internet. Sousveilling individuals now can invert an organization's gaze and watch the watchers by collecting data on them.

The development of wearable computing fits well with contemporary transformations of the social ecology. While surveillance is a manifestation of the industrial and postindustrial eras of large hierarchical organizations that have efficiently employed technologies in neopanopticons of social control, there is now a turn from such organizations to "networked societies" (Wellman, "Physical Place and Cyberspace"; Wellman, "The Network Community"; Castells). Rather than being embedded in single communities or work groups, individuals switch among multiple, partial communities and work teams. They move about, both socially and physically. Where centralized mainframe computers served the needs of large hierarchical organizations, personal computers better fit the needs of people in networked organizations and communities who move with some autonomy among geographically and socially dispersed work teams, friends, and activities. Yet personal computers are still rooted to desktops at the office and tabletops at home. They are still wired into computer networks. Wearable, wireless computers better fit the needs of people to be physically mobile as they move between interactions with workmates and community members. As the developed world transforms from small-group to person-to-person interactions, wearable, wireless computers are a tool for personal empowerment.

We describe here an attempt to use newly invented forms of wearable computing (Mann, "Wearable Computing"; Mann and Niedzvieki) to empower individuals in at least some aspects of their encounters with organizations. These inventions call into question Aldous Huxley's assertion that "technological progress has hurt the Little Man and helped the Big Man" (43). We examine how using wearable computing devices can promote personal empowerment in human technology and human interactions (Mann, "Wearable Computing"; Fogg). Two key issues are the extent to which organizational surveillance can be challenged and the ways that organizations respond to such challenges. We describe and analyze here a set of performances that follow Harold Garfinkel's ethnomethodological approach to breaching norms. We gain insight into these norms by deliberately not acquiescing in surveillance and by performing visible and explicit sousveillance. By breaking organizational policies, these performances expose hitherto discreet, implicit, and unquestioned acts of organizational surveillance.

Like Gary Marx, we are interested in how "the relationship between the data collector and the subject may condition evaluations, as may the place" ("An ethics for the new surveillance" 176). We attempt, as a systems analyst might, to engage our points of contact (e.g., managers, clerks, security workers) without claiming to understand complicated internal hierarchical considerations or politics within large bureaucratic, sometimes multinational, organizations. Instead, when the performers instigate situations in order to gauge the degree to which customer service personnel will try to suppress photography in locations where it is forbidden, they break unstated rules of asymmetric surveillance using new wearable computing inventions (Mann and Niedzviecki). Hence, a key goal of the performances reported here is to engage in dialogues with front-line officials and customer service personnel at the point of contact in semipublic and commercial locations.

Collecting digital images via photographs or videos is usually prohibited by store personnel because of stated policies and either explicit or unconscious norms, or unconscious norms that are only realized when they are breached. The surveilled become sousveillers who engage social controllers (e.g., customs officials, shopkeepers, customer service personnel, security guards) by using devices that mirror those used by these social controllers.

Uncertainty surrounds these performances; no one is ever sure of the outcome of the interaction between device, wearer, and participants. Design factors can influence performances: the wearing of technology can be seen by participants as either empowering or threatening, depending on the type of technology, the location, and how it is presented and represented. For example, people who use familiar mobile devices such as laptop computers and PDAs are perceived as more socially desirable than those with less familiar devices such as wearable computers and hands-free mobile phones (Dryer et al.).

The five performances described in this chapter were designed to show people could actively use sousveillance-enabling wearable computing devices. Each performance responds to different situations in which sousveillance techniques can be used to explore surveillance situations. The performances—held in streets, shops, restaurants, shopping malls, and department stores in a large Canadian city—range from situations in which passers-by are shown how they may passively become the subjects of observation to situations in which sousveillance, using covert and overt wearable computing devices, engages organizational surveillance.

In interactions among ordinary citizens being photographed or otherwise having their image recorded by other apparently ordinary citizens, those being photographed generally will not object when they can see both the image and the image capture device (see Performance 1) in the context of a performance space. This condition, where peers can see both the recording and the presentation of the images, is neither "surveillance" nor "sousveillance." We term such observation between equals "*co*veillance," an example of which could include one citizen watching another.

In conditions of interactions among ordinary people, those being coveilled generally will not object when they can see images being recorded from a concealed image capture device displayed on a wearable device that is part of a performance space (see Performance 2). By contrast, organizations engaged in surveillance generally will object to people engaging in obvious sousveillance in their establishments (see Performances 2 and 3). Surveillers will object more to the social act of challenging their authority through sousveillance than to the actual existence of sousveillance (see Performances 3 and 4). The objections that surveillers have with sousveillance can often be overcome by promoting the sousveillance to a high-level coveillance. Such high-level coveillance consists of essentially one large corporation monitoring another large corporation in the establishment where the performance takes place (see Performance 5). An example of this might be when security surveillance in a shopping mall overlaps an individual store's surveillance where a sense of mutual "safety" is acknowledged.

Performance 1: A Wearable Computer with a Data Projection System

This performance takes place on a public street and involves a wearable computer, a high-power mercury vapor arc lamp, and a data projector running from a backpack-based 120-volt battery (see Figures 2 and 3). The projector is aimed at the ground, with an image projected right-side-up to people facing the wearer. At this stage, the wearer of the device walks through the crowded downtown streets of a major metropolitan city on busy evenings. This performance is designed to gauge the reactions of ordinary citizens toward the device itself, unaccompanied by any explicit breach of actual or implicit rules or regulations.

When the devices are inactive, the wearer of the devices and the device itself are not foci of attention, although passers-by approach the computer wearer and ask questions unrelated to the project. For example, some ask for directions to nearby places (as if the wearer might have access to online data). Once the setup is complete, the display stimulus consists of the dynamic video of passers-by combined with the text caption "www.existech.com" projected on the ground.

The nature of the displayed material affects attitudes toward and perceptions of the device itself. For example, when the text displayed on the ground contains a ".com" URL, many people associate the device with a corporation. They approach the wearer of the device, asking questions such as "what are you selling today?" The commercial nature of

a.   b.

Figure 2a.   The wearable device contains a 1 GHz P3 CPU, a rendering engine, and a high-power mercury vapor arc lamp data projector within a black, flame-retardant Nomex uniform custom-tailored to fit the wearer. Here a person can see his or her own image together with other computer-generated material.

Figure 2b.   Close-up view showing the output of the high-intensity data projection system

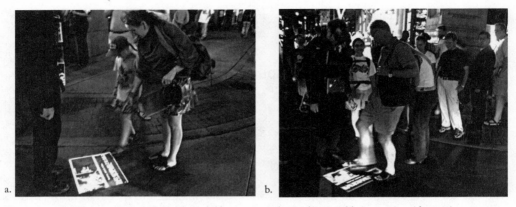

Figure 3a.   On the street, people bring their children over to play in the wearable interactive video environment and performance space.

Figure 3b.   Large crowds gathered to see the interactive environment. Even adults enjoyed playing in the interactive space.

the Web address contextualizes the device and its wearer as a marketing tool. This fits within an often expected and accepted use of public space. Experience gained from this type of performance suggests that the level of tolerance and acceptability toward the device and wearer relates to how it is contextualized within the existing knowledge and experience of people who encounter it. If the device appears to be sanctioned by a corporation or some other credible external authority, the level of acceptance is high and the technology itself is seen as a form of authority. Potentially critical audiences, such as shoppers or young adults lining up for fashionable dance clubs, were positive toward product displays on the device but negative toward artistic or satirical displays.

In short, when the performance is done in public spaces and appears to be organizationally related, acceptance by the public appears to be high. Surprisingly, people approved of the new form of advertisement in which live images were captured and rendered into computer-generated ads that included the subjects as models. People rarely object to their images being used in marketing.

Performance 2: Projected Data with Input from a Hidden Camera

This performance makes the source of the projected images less visible than in Performance 1. The use of the same highly visible projection, but with a hidden camera, sets up a disconcerting discrepancy in expectations between the technology used to capture an image and the projection of that image. A concealed infrared night vision camera is used to capture live video of passers-by. In the simplest form, the live video output of the hidden camera is displayed directly to the data projector. The effect of the hidden video camera remains obvious by virtue of the intense beam of the data projector and the arrangement of the projection.

In other forms of this performance, text, graphics, and other content containing images from the hidden camera are integrated on-the-fly and rendered to the data projector for the audience. Provocative text messages such as "ADVERTISING IS THEFT of personal solitude" are mixed with video from the concealed night vision camera system (see Figure 2).

A common reaction is that people try to find the hidden camera. They appear captivated (and sometimes amused) by its apparent physical absence despite its obvious functionality. Various text, graphics, and other subject matter—mixed in with live data, and displayed by the wearable data projector—evoke diverse responses. The most visceral responses occur when people see their own picture incorporated into the display. For example, when images of people are captured and then turned into a computer-generated advertisement, people pay more attention to the advertisement in which they are the subjects than they do to other, similar video material. People immediately recognize the appropriation of their image by a concealed, and therefore disconcerting, means.

The system gives rise to a roving interactive performance space where the roles of performer versus spectator, as well as architecture versus occupant, are challenged and inverted (see Figure 3). Passers-by become street performers and artists on the wearable stage that reflects their images to them. The stage itself, ordinarily thought of as a piece of architecture, has become a piece of clothing. Of course, the ability to choose to participate mitigates the invasiveness of the situation.

These relationships become more complex when wearing the device into spaces such as shopping malls that are semipublic rather than fully public. The potential for confrontation between the wearer and security personnel increases by moving into the more highly surveilled spaces of malls and stores while wearing the hidden camera and the projector. The device also loses much of its playfulness as it moves across this invisible border. Therefore, the more highly surveilled a space is, the more objections are raised about such sousveillance, regardless of whether the content displayed is viewed as satire or as an advertisement.

Performance 3: Making the Camera Obvious

Two cameras are used with the high-intensity wearable projection computer devices: the concealed infrared night vision camera of Performance 2 and an additional digital camera of the ordinary consumer variety that has been head-mounted. The purpose of using the additional camera is to make the act of taking a picture obvious. The additional camera chosen, a Kodak DC 260, looks like a traditional camera. It has a loud click sound (synthesized by its built-in speaker so that it sounds like a film camera) and a built-in electronic flash that calls attention to itself whenever it takes a picture.

When people turn to see what caused the flash, they see their pictures projected on the ground. To make the image capture more obvious, both pictures (freeze-frame stills and live video) are displayed side-by-side. The flash serves to clearly indicate that a picture is being taken every nineteen seconds (the update rate of the still camera). Text such as "CAMERAS REDUCE CRIME . . ." is used in the projection, together with the still and video displays (see Figure 4).

During Performance 3, social controllers often object to the taking of pictures because of organizational policies against sousveillance. However, the situation changes when the camera wearer attributes the acts of photographic data collection to external circumstances or to the camera wearer's apparent lack of control over picture taking. Various externalizers are used in the performance:

Figure 4.   Projected text: "Cameras reduce crime; for your protection your image shall be recorded and transmitted to the EXISTech.com image and face recognition facility."

1. The wearable computer system is completely hands-free. The wearer has no controls, no keypads, no mice, no buttons to push, and no other form of control over the device.
2. The device is automated or controlled externally so that it continues to take pictures while the wearer is explaining to the surveiller that it is beyond his or her control.

3. The wearer appears unable to remove the device. For example, the wearer explains to the surveiller that the device is held on by screws for security purposes. In this case, a skull frame with dermaplants and comfort bands are screwed to the eyeglass frames so that the wearer cannot remove the device. Other variations on externalization themes include deliberately modifying the camera ahead of time so that it "malfunctions" and gets stuck in the "on" position.

In addition to these physical externalities, the wearers create social externalities that suggest that they are required to wear the device due to various external obligations such as that the wearer is bound by contractual obligation to take pictures or that the wearer's livelihood depends on doing this. When "malfunctions" occur, the same types of social controllers—shopkeepers, customer service personnel, security officials, and so on—accept the fact that the wearers are taking pictures in their establishments.

The greater the appearance that the sousveiller has personal control over the device, the less acceptable the act of sousveillance becomes. For example, the level of tolerance and acceptability for taking pictures varies according to the degree of a "will not/may not/cannot" externality continuum. If the wearers explain that they are not in control of the devices and do not know when the devices take pictures, then the majority of surveillance personnel do not object to wearable devices. Surveillance personnel may initially object to the photography, but if the wearers of recording devices can show that they are not in control of the technology they are wearing, surveillance personnel are often mollified.

In other situations, if the sousveillance wearer appears to be "just following orders" of an external authority, thus mirroring the usual response patterns observed in surveillance personnel, the act of taking pictures is tolerated. Such externalization was made famous by Hannah Arendt's "Eichmann defense." The performers use a wearable camera whose use is made obvious by a flash and a loud click for each picture that is followed by a display of the picture. This produces a negative reaction when used without any attribution to external sousveillance authority. But this negative reaction often disappears when the picture-taker concomitantly uses a headset with microphone and says loudly to a remote "boss": "They seem to be objecting to having their pictures taken." The sousveillance wearer's apparent compliance with a credible external authority reduces objections made by surveillance personnel in a manner similar to Stanley Milgram's discoveries of obedience to authority.

Performance 4: Sousveillers Presenting Pictures of a Site to the Surveillers

The same kind of surveillance domes used by establishments can be used in wearable computing performances (Mann and Niedzviecki) (see Figure 5). These performances use wine-dark hemispheres similar to the seemingly opaque domes commonly found on the ceilings of stores. The fact that the domes may or may not contain cameras creates an important design element for the wearers because it is possible to arrange situations such that wearers do not know if their devices contain a camera. If questioned about the wearable domes, wearer would be able to reply that they are unsure what the dome contains.

Video recordings used in Performance 4 had been made prior to the performance by entering the shops with hidden cameras and asking various surveillance personnel what the domes on the ceilings of their shops were. In one case, customer service personnel

a.

Item #10034
IR NecklaceDome
$595 US

Item #23166
DoubleWhammy
$1490 US

Item #18451
SatchelDome 5
$795 US

Item #12199
AstroBoy
$795 US

Item #22267
ShootingBack NetPack
$849 US

above – Item #14155
WaistDome
$495 US

below – Item #19975
HeartCam
$1495 US

b.

Figure 5.   EXISTech's corporate brochure was created to present the artifacts of sousveillance in the context of high fashion as high-cost purchased goods. This contextualizes sousveillance products and serves as a legitimate element of conspicuous consumption in our consumerist society.

explained that the domes on their ceiling were temperature sensors. In another situation, a record store owner asserted that the store's dark ceiling domes were light fixtures. By using flat panel displays to play back the recording to the customer service personnel, their surveillance is reflected back to them as sousveillance.

In practice, surveillance personnel's appeal to authority can be countered by the sousveillers appealing to conflicting authorities. To be most effective, the sousveilling camera or projector wearer needs to be operating under social control policies in the same way that the surveillance worker or official is operating under company policies about surveillance. In this way, the wearer and the employee acknowledge each other's state of subordination to policies that require them to photograph each other. While the wearer and the employee engage in what would normally be a hostile act of photographing each other, they can be collegially human to one another and discuss the weather, sports, and working conditions.

Performance 5: Conspicuously Concealed Cameras

Whereas previous performances encountered resistance from certain surveillance establishments such as pawnbrokers, jewelry stores, and gambling casinos, the goal of Performance 5 is to create ambiguous situations in which wearable data-gathering devices are conspicuously concealed (Mann and Guerra). In this example, blatantly covert domes are used, together with a high-quality brochure that corporatizes and commercializes the tools of sousveillance. Figure 4 shows a line of products, along with the corporate brochure that was created to present the artifacts in the context of purchased goods. Store employees objecting to the wearing of such devices would also, by implication, be objecting to products of the consumerist society they are supposed to be upholding.

The wearable computers with domes evoke dialogue that varies as the size of the dome varies. For example, in one performance, a series of people entered an establishment wearing progressively larger domes until a complaint was raised. In some performances, performers played back video recordings of the same customer service personnel or of customer service personnel in other shops.

Conclusions: Sousveillance in Society

The performances show how certain kinds of rule violation can be deliberately used to engender a new kind of ecological balance in public and semipublic spaces. They show public acceptance of being videotaped as an act of surveillance in public places. When such data collection is done by ordinary people (such as the performers) to other ordinary people, it is often accepted. However, when data projectors show surveillance officials the data that has been collected about them, there is less acceptance. Organizational personnel responsible for surveillance generally do not accept sousveillance from ordinary people, even when data displays reveal what the sousveillers are recording. The only instances of acceptance are in Performance 4, when surveiller and sousveiller can find common ground in both doing "coveillance" work for symmetrically distant organizations.

The performances described here engage, challenge, and invert the power structure of networked surveillance. The role reversal between the surveilled individual and those performing the surveillance allows for the exploration of the social interactions that are generated by these performances. It raises questions for further inquiry, primarily the issues of collective empowerment and self-empowerment within the panopticon of social surveillance and the governance of public and semipublic places (Foucault, *Discipline and Punish*; Ostrom). Also, the performances show how the public can bring the technologies of surveillance to bear on surveillance workers whose profession it is to maintain such hierarchies of control. There is an explicit in-your-face attitude in the inversion of surveillance techniques that draws from the women's rights movement, aspects of the civil rights movement, action research, and radical environmentalism. Thus, sousveillance is situated in the larger context of democratic social responsibility.

Surveillance cameras threaten autonomy. Shrouding cameras behind a bureaucracy results in somewhat grudging acceptance of their existence in order to participate in public activities such as shopping, accessing government services, and traveling. By having this permanent record of the situation beyond the transaction, social control is enhanced. Acts of sousveillance redirect an establishment's mechanisms and technologies of surveillance back on the establishment. Sousveillance disrupts the power relationship of surveillance when it restores a traditional balance that the institutionalization of Bentham's Panopticon had disrupted more than a century ago. Sousveillance is a conceptual model of reflective awareness that seeks to problematize social interactions and factors of contemporary life. It is a model with its root in previous emancipatory movements and with the goal of social engagement and dialogue.

The social aspect of self-empowerment suggests that sousveillance is an act of liberation, of staking our public territory, and of leveling the surveillance playing field. Yet the ubiquitous total surveillance that sousveillance now affords can be seen as either an ultimate act of acquiescence on the part of the individual or a return to the metaphoric village in which everyone knows what everyone else is doing, even if it is now within Marshall McLuhan's global village.

Universal surveillance and sousveillance may, in the end, only serve the ends of the existing dominant power structure. Universal surveillance and sousveillance may support the power structures by fostering broad accessibility of monitoring and ubiquitous data collection. Or, as William Gibson comments in the documentary film *Cyberman*, "You're surveilling the surveillance. And if everyone were surveilling the surveillance, the surveillance would be neutralized. It would be unnecessary."

In such a coveillance society, the actions of all may, in theory, be observable and accountable to all. The issue, however, is not about how much surveillance and sousveillance is present in a situation but how it generates an awareness of the disempowering nature of surveillance, its overwhelming presence in Western societies, and the complacency of all participants toward this presence.

Despite police espousal of "neighborhood watch" programs, few of us live in a world where watching one's neighbors is a practical mode of social control. Such close local observation is mostly found in preindustrial societies, their remnants in rural and urban villages, and in specialized situations (Ostrom). Urban houses are often vacant while families are scattered about at various activities. Most friends and relatives live in other parts of the city, continent, or globe. Many coworkers are not collocated in the same spaces, and most shopkeepers do not know their customers personally.

Surveillance is a manifestation of the industrial and postindustrial eras, of large hierarchical organizations efficiently employing technologies in neopanopticons of social control. However, in contemporary networked societies, individuals switch among multiple, partial communities and work teams rather than being embedded in single communities or workgroups (Wellman, "The Network Community"). Accordingly, in networked societies, people are more likely to want sousveillance and coveillance, for they lack the protection of the village/community or hierarchical organization. A potential solution to fill the gap between community or institution and networked individualism can be found in newly developed technologies that allow individuals to surveil the surveillers. Camera phones are in the hands of children who can upload images on-the-fly to image-sharing

Web sites such as YouTube and Flickr. (We wonder if there is a children's video on YouTube sousveilling parental spanking.) Flash mobs can coalesce around a specific issue, and the activists can melt away just as quickly as they appeared. By enabling people to be simultaneously the master and the subject of the gaze, wearable computing devices offer a new set of strategies and voices in the usually one-sided dialogue of surveillance.

Works Cited

Arendt, Hannah. *Eichmann in Jerusalem: A Report on the Banality of Evil.* New York: Viking, 1963.

Bair, Deidre. *Samuel Beckett: A Biography.* New York. Simon and Schuster, 1978.

Bentham, Jeremy. *The Collected Works.* London: Athlone Press, 1968.

Cannon, Lou. *Official Negligence: How Rodney King and the Riots Changed Los Angeles and the LAPD.* Boulder, Colo.: Westview, 1999.

Castells, Manuel. *The Rise of the Network Society.* Rev. ed. Oxford: Blackwell, 2000.

Cyberman. Dir. Peter Lynch CBC, 2001.

Dryer, D. Christopher, Chris Eisbach, and Wendy S. Ark. "At What Cost Pervasive? A Social Computing View of Mobile Systems." *IBM Systems Journal* 38.4 (1999): 652–76.

Fogg, B. J. "Captology: The Study of Computers as Persuasive Technologies." *Extended Abstracts of CHI '97.* Atlanta: ACM Press, 1997.

Foucault, Michel. *Discipline and Punish.* Trans. Alan Sheridan. New York: Vintage, 1977.

———. "Prison Talk." *Power/Knowledge: Selected Interviews and Other Writings 1972–1977.* Ed. Colin Gordon. New York: Pantheon, 1980. 37–54.

———. "Question of Geography." *Power/Knowledge: Selected Interviews and Other Writings 1972–1977.* Ed. Colin Gordon. New York: Pantheon, 1980. 61–77.

Garfinkel, Harold. *Studies in Ethnomethodology.* Cambridge: Polity, 1967.

Huxley, Aldous. *Brave New World Revisited.* New York: Harper, 1958.

Ingram, Rick E. *Privacy and Psychology.* New York: John Wiley and Sons, 1978. 35–57.

Lefebvre, Henri. *The Production of Space.* Trans. Donald Nicholson-Smith. Oxford: Blackwell, 1997.

Lovink, Geert, and Florian Schneider. "A Virtual World is Possible: From Tactical Media to Digital Multitudes." *Artnodes: Interseccions Entre Arts, Ciències I Technologies.* Barcelona: FUOC, 2003.

Mann, Steve. *Intelligent Image Processing.* Mississauga: John Wiley and Sons, 2001.

———. "'Reflectionism' and 'Diffusionism': New Tactics for Deconstructing the Video Surveillance Superhighway." *Leonardo* 31.2 (1998): 93–102.

———. "Wearable Computing: A First Step toward Personal Imaging." *IEEE Computer* 30.2 (1997): 25–32.

Mann, Steve, and Robert Guerra. "The Witnessential Net." *Proceedings of the IEEE International Symposium on Wearable Computing* Oct. 2001: 47–54.

Mann, Steve, and Hal Niedzviecki. *Cyborg: Digital Destiny and Human Possibility in the Age of the Wearable Computer.* Toronto: Random House Doubleday, 2001.

Marx, Gary T. "The Engineering of Social Control: The Search for the Silver Bullet." *Crimes and Inequality.* Eds. John Hagan and Ruth Peterson. Stanford, Calif.: Stanford University Press, 1995. 225–35.

———. "An ethics for the new surveillance." *The Information Society* 14 (3) (1998): 171–85.

———. "A Tack in the Shoe: Neutralizing and Resisting the New Surveillance." *Journal of Social Issues* 59.2 (2003): 369–90.

Milgram, Stanley. *Obedience to Authority.* London: Tavistock, 1974.

Ostrom, Elinor. *Governing the Commons: The Evolution of Institutions for Collective Action.* Cambridge: Cambridge University Press, 1990.

The Prisoner. By Patrick McGoohan and David Tomblin. London: Everyman Films and ITC, 1967.

Rhodes, Bradley, Minar Nelson, and Josh Weaver. "Wearable Computing Meets Ubiquitous Computing: Reaping the Best of Both Worlds." *Proceedings of The Third International Symposium on Wearable Computers.* ISWC: San Francisco, 1999. 141–49.

Rogers, Ted W. "Disrupted Borders: An Intervention." *Definitions of Boundaries.* Ed. Sunil Gupta. London: Rivers Oram, 1993.

Stanley, Jay, and Barry Steinhardt. "Bigger Monster, Weaker Chains: The Growth of an American Surveillance Society." Washington, DC: Technology and Liberty Program, American Civil Liberties Union, January 2003. http://www.aclu.org/privacy/gen/15162pub20030115.html.

Ward, Tom. "The Situationists Reconsidered." *Cultures in Contention*. Eds. Douglas Kahn and Diane Neumaier. Seattle: The Real Comet, 1985.

Webster, C. William R., and John Hood. "Surveillance in the Community: Community Development Through the Use of Closed Circuit Television." *Community Informatics: Shaping Computer-Mediated Social Relations*. Eds. Leigh Keebler and Brian Loader. Routledge: London, 2001. 220–39.

Wellman Barry. "The Network Community." *Networks in the Global Village*. Ed. B. Wellman. Boulder, Colo.: Westview, 1999. 1–48.

———. "Physical Place and Cyberspace: The Rise of Personalized Networks." *International Journal of Urban and Regional Research* 25.2 (2001): 227–52.

13

Ambient Video: The Transformation of the Domestic Cinematic Experience

Jim Bizzocchi

If you are standing five feet away from a four-foot-wide, high-definition video screen, is it television or is it IMAX? Or is it something else entirely?

The yule log burns cheerfully in the fireplace. Or does it? In 1996, I created a visual conversation piece for our annual Christmas party. I installed a small charcoal-grey TV set in the fireplace, and ran prerecorded footage of a burning log, which I had shot in the very same fireplace. The illusion was interesting enough to enthrall our guests in those moments when they had run out of immediate conversation. This video installation was a form of ambient media in the tradition of Brian Eno's ambient music, which "must be able to accommodate many levels of listening attention without enforcing one in particular; it must be as ignorable as it is interesting" (Eno).

My burning log video certainly fit both of Eno's criteria: it had the capacity to be both interesting and ignorable. I got a fair bit of pleasure from it, and I was proud of my little piece of domestic video art. However, my pride was taken down a notch when I discussed the matter with a senior colleague, who informed me that the burning log was a video cliché. He had done one himself a long time ago, and many other versions have appeared since that time. Tapes of a burning log (sometimes called a "Yule log") have been available commercially for years, and you can now find a variety of DVD versions on the Internet. My cable company's community access channel has been broadcasting their burning Yule log video every Christmas for years. In addition, there are now three *other* channels on my cable service that feature their own holiday log. What may be the oldest broadcast version was first aired on New York's WPIX in 1966. The WPIX log has won its local time sweeps for the past three years, is now available on the station's digital channel, and it reaches sixty-five million American homes through WPIX's corporate stablemate, superstation WGN. The Yule log phenomenon is a global one; Reuters reports that a German woman saw her local version of the Yule log and called the fire department!

The various electronic avatars of the Yule log—VHS, DVD, cablecast, broadcast, homemade—have burned away in millions and millions of homes for decades.[1] Unlike our normal conception of the televisual, this version of video imagery is a truly ambient experience. In my house, as in countless others, the various versions of the log play in and around the background of our lives. From time to time, the dancing electronic flames will capture, and even hold, the attention of one or two people in the household. Inevitably, the moment of concentrated gaze passes, and attention is shifted to another activity or to more immediately engaging video material. Despite our inconsistent attention, the log itself, as with millennia of campfires and fireplaces, maintains its role in our visual environment. The question for us is whether the Yule log is merely an interesting but strictly limited holiday phenomenon or whether it is a harbinger of a more widespread cultural trend toward a new type of ambient video form.

Reception, Experience, and Production Practice

In order to answer this question, consider some of the implications of the forms of digital video production and presentation tools that have recently emerged. Emergent forms of mediated experience carry within themselves fresh aesthetic opportunities. This McLuhanesque call for media specificity is not an argument for a simplistic form of technological determinism as derided by Raymond Williams. The reality is more complicated than that. As artists and creators work within a new medium, its effective poetics are revealed through practice and experimentation. In technologically based art, these poetics are refined through interconnected dialectics of art, commerce, and critical discourse. This dialogue between the creative and the critical is equally important for the development of new forms of contemporary video expression.

The initial visual poetics of video were derived from those of film. However, the two were never identical; there were critical differences that led to variance in the production practices and the effective poetics of the two media. This chapter is concerned with two of these differences. One is the difference in visual quality—in particular, scale and resolution. The large, rich, finely textured visuals of theatrical film (or even well-crafted 16 mm film footage) are far superior to the truly marginal quality of standard television images. The second difference lies in the conditions of reception. Theatrical film is seen in a magic black box, a glowing shrine to the suspension of disbelief. Television and video are typically seen in the home, where the entertainment appliance vies for our attention along with the telephone, the refrigerator, the washroom, and the daily distractions and companions of our everyday lives.

One of these two differing conditions will shift dramatically, the other is harder to predict. The condition that will change is the visual quality of the experience. Video capture and display technologies are rapidly improving. More difficult to anticipate and summarize are the environmental parameters of the home video experience, to which we will return later in this chapter.

The Evolution of the Video Image

The changes in the visual quality of video are relatively predictable. The family of television appliances has undergone a significant visual upgrade. Picture size gets bigger and bigger and picture quality gets better and better. The size trend has been a steady growth. The quality trend has been punctuated by advances in video playback and distribution technology such as cablecasting, laser discs, satellite distribution, DVD, advanced consumer video recording capability, and digital multicasting. Unfortunately, with few exceptions, the current quality of these formats is bound by the overall limitations of consumer television. The engineer's lament that NTSC stands for "Never Twice the Same Color" has the ring of sad truth for those that love a reliable and crisp image. PAL and SECAM are certainly improvements on the North American NTSC standard, but they will never rival cinema for visual quality or impact.[2]

The quality and impact of the home video experience is now making a double quantum jump. The first is the introduction of high-definition television standards for broadcasters and producers of consumer electronics. The second is the increasing size and the decreasing price of flat-panel display screens. The obtrusive box in the corner with the marginal picture is becoming an elegant (and large) frame on the wall, presenting imagery that is closer to cinematic standards than anything in our previous television experience.

The commercial momentum of this change is considerable. Evidence is provided by a review of newspaper advertisements of home video equipment and confirmed through sales statistics and projections (Joseph and Fasold; Kitadata and Takahashi). As picture size grows, standard television sets are being steadily supplanted by quasi-flat projection television, and true flat-panel (plasma and LCD) video display. The new receiver-monitors in all configurations include "HDTV" (high definition television) or "HD-compatible" as part of their marketing pitch. The wide-screen, high-definition experience is being sold hard with a reliance on movies, sports, and lifestyle as the marketing drivers. The HD marketing pitch is being reinforced in several new directions. Sports are a big sell for the men in American households. "Action so real you will want to wear a helmet," reads the sports-oriented ad for one big-screen, HD, flat-panel television ("Action So Real . . . "). This may be hyperbole, but the big, high-resolution screen has the capability to solve a critical problem in televising team sports: how to reconcile the need to show the flow of the entire play and at the same time maintain narrative identification with the individual stars. The new screen technologies can offer both in one large, high-resolution wide shot. For some sports, like hockey, tennis, and golf, these HD screens offer the further opportunity to actually see the ball or the puck during play!

The marketplace has a different television hook for youth and adolescents. Two of the current generation video game platforms (Microsoft's Xbox 360 and Sony's PlayStation 3) feature high-definition video output. The big screens in millions of living rooms and family rooms will amplify the performative aspect of the electronic game experience. Completing the domestic loop, women (who are key to the purchasing of home electronics) are increasingly drawn to the style and design beauty of the flat-panel devices (Pearce). A review of the photo spreads in home sections of newspapers reveals the increasing inclusion of a visual "triple-play" in the depictions of the perfect living room: fireplace, picture window, and large, flat-panel television hanging on the wall. All three are devices that bring ambient visual pleasure into our day-to-day lives. In any case, the stereotypical nuclear family is being tempted on all fronts—husband, wife, and children—by the lure of HD technology. Until now, high comparative costs confined this item to the early adopter end of the technology acquisition spectrum, with the projection TV playing a role as a less expensive "starter" big set; however, there is a logic to the adoption curve for the flat-panel video units. As HDTV distribution continues to grow and more consumers are ready to move up from projection boxes and traditional picture tubes, flat-panel technology development costs are amortized over longer and larger production runs, and prices for the wall units are inevitably coming down.[3] At the same time, we will see continued development of the next generation of flat-panel technologies, such as OLED (organic light-emitting diode) and HDR (High Dynamic Range) displays.

The introduction of high-definition, flat-panel displays creates unprecedented conditions of televisual reception experience. For the first time, cinematic-quality visuals are situated in our homes. The effects of this development will be amplified by a variety of other digital production and post-production tools. Inexpensive, high-quality digital video cameras provide widespread opportunities for experimentation with new forms of video expression. The latest version of prosumer cameras offer HD-quality visuals for approximately $3,000. These high-resolution images can be edited, processed, and transformed with an array of sophisticated post-production software. Final Cut Pro, Adobe Premier, After Effects, and related tools offer thousands of digital filmmakers opportunities for layering, segmentation, combination, metamorphosis, and transformation that were formerly confined to the most costly studio operations or the most obsessed of film artists.

Implications for Video Content

These changes in video quality and capability carry implications for video production and the televisual experience. The first is a return to a more film-like aesthetic. The starting point is the recovery of a robust spatial representation. Television imposed severe limits on the treatment of scale and perspective. The loss of cinematic image size and resolution was a double whammy for the visual impact of the original televised picture. The long shot lost its expressive and its communicative powers, and the close-up became privileged to the point of imperative.

The new display technologies reverse that trend. The scope of the reversal will depend on questions of screen size and resolution, but the trend will be to make video much more film-like in its presentation characteristics and therefore in its production aesthetics. In fact, the combination of size, resolution, and viewing distance may eventually bring the reception conditions of home video closer to Cinerama than to conventional movie formats. The relevant question will be, "If you are standing five feet away from a ten-foot-wide high-definition video screen, is it television or is it IMAX?"

Even before this extreme evolution, the new video form will differ from the old video form in many of its fundamental poetics. As visual field, image size, and resolution approach cinematic standards, the wide shot will be reprivileged, and the close-up will become far less critical. In some situations, the use of tight close-ups will become counterproductive.

This change in treatment of subject scale will support a new freedom for the choice of editing pace. Television's devaluation of the wide shot lent an impetus to faster cutting for visual storytelling. Classic cinematic composition in depth was a form of spatial montage. For filmmakers such as Orson Welles or John Ford, narrative detail could be arranged within a long single shot and successively privileged through sound, lighting, and blocking of action. Television was perfectly capable of using the long (and wide) take to support dialogue, but it needed a different strategy for visual storytelling. Its reliance on medium and close shots necessitated the sequencing of any critical visual narrative elements. Story tended to be supported through a succession of tighter images rather than through the visual dynamics of a single rich image. The height of this effect was exhibited in several subgenres unique to television: the commercial, the series opening signature sequence, and the music video. These forms faced a unique set of constraints. Not only did they have to contend with the visual limitations of standard television, they also had to face the double test of working well upon first viewing yet standing up to repeated examination. One of their defining tactics was to push the limits of temporal montage, increasing the cutting pace enormously. Their joint effect on the poetics of the moving image was far-reaching indeed. The video "short form" triumphed in its own right and, in turn, affected the poetics of longer television shows and of mainstream cinema.

As a result of our exposure to the fast-paced video short form, our ability to take in visual information has increased tremendously (Stephens 154). However, temporal acceleration is not the only path to a rich visual information environment. One has to consider the effect of the new display standards on the fundamental poetics of the medium. Lev Manovich is attuned to the implications of the evolutionary nature of the screen. He recognizes that monitors are getting bigger and will eventually become wall-size (114–15). Having established this context, he points out that "spatial montage represents an alternative to traditional cinematic temporal montage" (Manovich 322). Manovich feels he is extending Eisenstein's conceptions of montage as an ongoing dialectic within a full range of audio-visual and spatial-temporal possibilities (Eisenstein). At the same time, Manovich relies on the role that digital technology has played in empowering creators. Digital art lends itself to fragmentation into parts and recombination into new and layered dynamic constellations. This potential gives video artists powerful tools to wield on their improved electronic palettes.

Two of these tools are the split screen and the layered transition. At the risk of a bad pun, the split screen has a checkered cinematic history. Its full capabilities have never been consistently exploited. Any one of us can name a few feature films that have used this technique: *The Thomas Crown Affair*, *The Boston Strangler*, *Woodstock*, and Abel Gance's *Napoléon*. Few of us could name as many as twenty examples in film's long history. In a similar vein, shot-and-scene transitions have been dominated by the hard cut, with minor attention to the lap dissolve, the fade, and a very small percentage of pattern wipes. More complicated transitions were possible, but the cost of optical effects in the film world and the lack of visual quality in the video world have limited their utilization.

Even given the mainstream cultural dominance of a relatively linear and unambiguous narrative tradition, the use of these multiformed visual devices has been low. However, the next several years may well test their aesthetic capabilities. The new video display units provide an appropriate platform, and contemporary digital forms provide the conceptual models. Jay Bolter and Richard Grusin point out that different media ceaselessly adapt and repurpose each other's forms and conventions. The windowed universe of the desktop and the Web is reflected in a rebirth of the fragmented-frame moving image environment. We see this effect in the news networks, in dramatic television series such as *24* and *Trial and Retribution* and in a number of contemporary films such as *The Rules of Attraction*,

Timecode, and *Run, Lola, Run*. At the same time, the morphing and collaging capabilities of digital post-production software supports a layered video experience that seamlessly blends varied backgrounds and subjects in a smooth temporal flow. The cinematic wipe flourished in the thirties and subsequently fell out of favor, but it and similar devices are undergoing a rebirth (Thompson and Bordwell). The beginnings of a revitalized aesthetic of layered transitions can be seen in television series such as *Home Improvement* and *Las Vegas* and hypermediated action-comic cinema such as *Hulk* and *Spider-Man*.

The renaissance of the scenic wide shot, the split screen, and layered transitions is indicative of a broader direction. The new screen technologies support and mandate a strong shift to the pictorial. Larger surface and higher resolution carry their own visual logic. Creators will inevitably exploit it, and viewers will come to expect it. Other pictorial directions will include an increased emphasis on lighting and composition, the hypnotic attraction of slow motion imagery, and the continued exploration of the moving camera. Long-form visual poems such as *Koyaanisqatsi* or *Baraka* are examples of a pictorial cinema that will help to define the aesthetic boundaries enabled through the new video formats.

Conditions of Reception

These opportunities are complicated by the situation of this rich visual field in a domestic consumer device. The question still stands: "Is the new video display television or is it IMAX? Or is it something else?" The key here is the question of foreground and background. Film is very much a foreground medium. We sit in a dark room, transfigured by the glowing image that dominates our visual world (supported by a sound experience as rich and as full as the visual). This is an environment completely adapted to the "willing suspension of disbelief" (Coleridge). Television, on the other hand, is a chameleon. It is capable of assuming either foreground or background status depending on several variables: the quality of the video experience, the exigencies of domestic life, and shifting user preference in any given moment.

Surprisingly few television critics address this aspect of the home television experience. Williams correctly identified "flow" as a powerful concept for the analysis and understanding of television, but Lynn Spiegel's introduction to the 1992 edition of Williams's seminal work *Television: Technology and Cultural Form* points out that the flow of programming is often interrupted or overridden by the flow of domestic life: "[Williams's methodology] didn't at all account for more everyday viewing procedures. It didn't account for someone preparing a sandwich, answering a phone, putting a child to bed—in short the flow of human activities that interact with the flow of television programs" (Spiegel xxvi). Television experience may be pleasurable precisely because it can be a casual activity. Jib Fowles sees selective attention as a basic characteristic of home television viewing that adds to the enjoyment of the experience. He cites a study of actual television usage which showed that 20 percent of the time when the television was on, there was no one in the room, and another 20 percent of the time, potential viewers were in the room but not paying attention to the screen. The rate of inattention grew to 50 percent during commercials (C. L. Allen, qtd. in Fowles). A robust understanding of the range of possible options for video production should take into account the variable attention actually paid to domestic television.

The new screen display formats approach the presentation quality of film but retain the figure-ground malleability of video. In combination, this describes a medium where there will be ongoing demand for foreground programming and, at the same time, some level of demand for background programming. We will still use the new screens to watch movies. The latest DVD (or its HD technical equivalent) will remain a domestic "destination event" that dominates our attention. At the same time, we will continue to use the device for standard television programming such as news, dramatic series, game shows, music, and sports. Our attention to these shows will vary tremendously, as it has for decades of television viewing.

Ambient Video

We are now drawing closer to the issues raised by the history of the burning Yule log video. Do the new digital display tools support the possibility for a new content direction—a true ambient video form? Will we see the development of a new class of domestic video art that will hang on our walls within these beautifully minimalist flat-panel video frames? The phrase "ambient video" would describe programming that is designed to run in the background but will sustain a certain amount of close attention at any time. A pejorative term for this type of programming is "video wallpaper." A more accurate and respectful term would be "video painting." The immediate digital antecedent is the computer screen saver; its aesthetic roots will be found in certain kinds of experimental cinema and video art. Its commercial beginnings can already be seen. The elemental companion to the fire of the burning log is the televisual liquid pleasure of the video aquarium. The DVD of *Finding Nemo* contains two different collections of video aquaria, one for standard video, the second for 16:9 format screens. A search of the Internet for "ambient video" reveals a range of companies selling ambient video DVDs. The visuals on these videos include fireplaces and aquaria as well as a wide variety of nature imagery: beaches, mountains, forests, and waterfalls. The sound tracks include a complementary range of background audio consistent with the ambient functionality of the visuals: classical music, choral arrangements, natural sound effects, and light electronica. Some of the advertising is aimed at specialized markets (VJs, religious institutions, and the wellness community), but much of it is clearly designed for more generalized domestic display. It is also clear that much of the appeal of this level of commercial product is to a kitsch instinct rather than substantive artistic appreciation.

Is the new ambient video doomed to this gimmicky level of creative endeavor? Is it merely a new standard for electronic velvet paintings? Luckily we can find precedent for substantive explorations into a new ambient video form in the history of video art and experimental film. Eno himself explored this form in the late '70s and throughout the '80s, producing ambient video works such as *Mistaken Memories of Mediaeval Manhattan* and *Thursday Afternoon*. Eno has plenty of company among artists working more directly with the moving image. Avant-garde filmmakers such as Andy Warhol (in his *Empire*, *Sleep*, and other works) explored the possibilities of the extremely long take. Michael Snow's *Wavelength* also explored the use of the slow shot and added his own sense of liminal narrative to Warhol's aggressively minimalist stance. Yoko Ono's *Sky TV* and *Apotheosis* (the latter with John Lennon) tied the long take to a pictorial sensibility and the play of landscape and skyscape. Bill Viola's works—from *The Reflecting Pool* to *Emergence*—combine a commitment to a luxurious screen time with an acute understanding of the painterly beauty of the carefully composed video image.

These and other works from the tradition of the innovative and experimental moving image demonstrate the potential for a new type of domestic televisual art. The prime characteristic for this type of programming is that it is pleasant, visually interesting, and capable of supporting occasional close viewing. It should change, but not too quickly, and the details of any particular change should not be critical over a limited time frame. This is ambient video—the "slow-form" reversal of forty years of intense development of the fast-paced television "short-form." Some work in this genre will be closely linked to a screen-saver aesthetic. This will include purely graphic abstract designs and geometrics, naturalistic motion graphics such as water and fire, and quasi-narrative artificial life environments. It will certainly include visual creations that are driven by music (such as the light shows built into Apple's iTunes and Microsoft's Windows Media Player).

Other work in this stream will be more cinematic. This variation will concentrate on rich and compelling visuals, making full use of the screen's size and resolution. Like the purely graphic screen-saver form, the aesthetic imperative for the cinematic version is visual ambience. The size and beauty of the visuals will capture a casual glance at any moment. The resolution and detail of the image will enable the subtle details that can sustain a more concentrated gaze. The incorporation of slow change and metamorphosis will support still longer and closer examination.[4] This form will privilege the use of nature sequences (fire, water, clouds, foliage, geology), slow motion, gradual transitions, visual effects, layered and convoluted imagery, and subtly embedded secondary visual artifacts.

The nuance of this direction will be the seduction of visual sensibility. The archetypal situation is a background visual during a cocktail party. People will converse and then glance at the screen during a pause in the discussion. The glance will be compelling for a moment, a minute, or several minutes. Then the conversation resumes, and the viewers withdraw their attention—until the next pause in their personal conversational flow. When the viewer is again ready, the screen will be there, revealing rich and living imagery at any given moment of choice.

It is worth noting that we are echoing the reception requirements of the video short form. Commercials, series openers, and music videos are designed to work on first viewing and to work on multiple viewings after that. Ambient video shares those difficult goals. It too must work immediately and sustain multiple viewings. However, there is a significant difference. The short forms are designed to compete for foreground attention in the contested reception environment of the home. The ambient video slow form does not contest. It waits. It is content to play in the background but always ready to assume foreground attention at the choice of the reader. Its capacity for repeated viewing cannot depend on temporal montage and fast pacing—these devices both require and command viewer attention. Instead, ambient video will be a more purely visual medium—relying on pictorial impact and the subtle manipulation of image, layer, flow, and transition. It will play on the walls of our homes, a window of infinite possibility capable of supporting any level of attention we care to bestow.

Notes

This chapter has been improved thanks to comments and insights from colleagues William Uricchio, David Thorburn, Pete Donaldson, Elizabeth Drew, Belgacem Ben Youssef, John Bowes, and Henry Jenkins. Earlier versions of this chapter have appeared in *Crossings: Electronic Journal of Art and Technology*, the proceedings of the 2003 International Conference on Entertainment Computing, and the *Journal of Moving Image Studies*. The author's research has been supported by the Canadian Social Science and Humanities Research Council, the Banff New Media Institute, and Simon Fraser University.

1.   There are a series of Web citations on the history of the WPIX log, the Reuters log-scare woman, and various Web-based sales sites for a video Yule log or virtual fireplace.

2.   The various international video standards reflect complex struggles over technological quality, control of commerce, and cultural hegemony. NTSC (National Television Standards Committee) is the North American standard and is often used in those countries most closely aligned with the American and Japanese marketing domains. PAL (Phase Alternating Line) was developed in Germany and adopted widely throughout the British Commonwealth and within other Western European spheres of economic and cultural influence. SECAM (Système électronique pour Couleur avec Mémoire) was developed in France and tends to be used in Francophone and former Soviet countries. (The cynical translation of SECAM is "Something Essentially Contrary to the American Method".) More germane to the argument in this paper is the fact that PAL and SECAM have better resolution (at least in the vertical dimension) than NTSC but do not approach the resolution quality of high-definition television standards.

3.   Of course, the saturation of the initial HD consumer market will be followed by further improvements in picture quality. For example, Scott Stevens points out that frame rate adds considerably to perceived resolution and visual impact, and proposed future generations of HDTV will undoubtedly include higher frame rates that will take advantage of this phenomenon (Stevens).

4.   Lev Manovich is pursuing a similar set of goals in his "soft Cinema" project. This groundbreaking work is designed to elicit a range of viewer responses that includes such modes as *glance, focus, observe, examine* and *study*. He includes a description of settings and architectures that complement the large screen and support the entire range of response intensities.

Works Cited

24. Fox Studios, 2001–2006.

"Action So Real . . ." A&B Sound commercial flyer. Vancouver, 26 Jan. 2005.

Apotheosis. By Yoko Ono and John Lennon. Film, 1966.

Baraka. Dir. R. Fricke. Samuel Goldwyn Company, 1992.

Bolter, Jay, and Richard Grusin. *Remediation.* Cambridge, Mass.: MIT Press, 1999.

The Boston Strangler. Dir. R. Fleisher. Twentieth Century Fox Film Corporation, 1968.

Coleridge, Samuel. *Biographia Literaria.* London: Clarendon Press, 1907.

Eisenstein, Sergei, *Film Form and Film Sense*, Cleveland OH, Meridian Books, 1957

Emergence. Dir. Bill Viola. Commissioned by the J. Paul Getty Museum, Los Angeles. Video, 2002.

Empire. Dir. Andy Warhol. Film, 1964.

Eno, Brian. Album liner notes. *Music for Airports.* EG/Virgin Records, 1979. 25 May 2004 http://music
.hyperreal.org/artists/brian_eno/MFA-txt.html.

Finding Nemo. Dir. Andrew Stanton and Lee Unkrich. Disney/Pixar, 2003.

Fowles, Jib. *Why Viewers Watch: A Reappraisal of Television's Effects.* Rev. ed. Newbury Park, Calif.: Sage, 1992.

Home Improvement. ABC/Disney, 1991–99.

Hulk. Dir. Ang Lee. Universal Studios, 2003.

Joseph, Jeff, and Lisa Fasold. "2003 U.S. Sales of Consumer Electronics to Hit New Record, Kissing $100
Billion, Says CEA." *Consumer Electronics Association Press Release.* 7 Jan. 2003. 5 July 2004 http://www.ce
.org/press_room/press_release_detail.asp?id=10138.

Kitadata, Mike, and Karl Takahashi. "Panasonic Steps Up New 'VIERA' Flat-Panel TV Series."

Koyaanisqatsi. Dir. G. Reggio. New Cinema, 1983.

Las Vegas. NBC, 2004–2006.

Masushita Electric Co. Ltd. Press Release. 10 May 2004. 5 July 2004 http://matsushita.co.jp/corp/news/
   official.data/data.dir/en040510-9/en040510-9.html.

Manovich, Lev. *The Language of New Media.* Cambridge, Mass.: MIT Press, 2001.

Mistaken Memories of Medieval Manhattan. Dir. Brian Eno. An OPAL Production, Toronto 1981 and
   London 1987.

Napoléon. Dir. Abel Gance. Société générale des films, 1927.

Pearce, Tralee. "TV Engineers Take Lessons from Interior Designers." *The Globe and Mail.* Toronto. 10 Feb.
   2005.

The Reflecting Pool. Dir. Bill Viola. Video, 1977–79.

Rules of Attraction. Dir. Roger Avary. Lions Gate Films, 2002.

Run, Lola, Run (Lola rennt). Dir. Tom Tykwer. Bavaria Film International and Sony Picture Classics, 1998.

Sky TV. By Yoko Ono. Video Sculpture, 1966.

Sleep. Dir. Andy Warhol. Film, 1963.

Spider-Man. Dir. Sam Raimi. Sony Pictures, 2002.

Spiegel, Lynn. Introduction. *Television: Technology and Cultural Form.* By Raymond Williams. 1974.
   Middletown, Conn.: Wesleyan University Press; Hanover, N.H.: University Press of New England, 1992.

Stephens, Mitchell. *The Rise of the Image, the Fall of the Word.* Oxford: Oxford University Press, 1998.

Stevens, S.M. "Time Traveler, Cinematic Theory, Perception, and Scientific Visualization" *Pixel, The
   Magazine of Visualization* 3, March/April 1992, 34–39.

The Thomas Crown Affair. Dir. Norman Jewison. United Artists, 1968.

Thompson, Kristin, and David Bordwell. *Film Art.* New York: McGraw-Hill, 1994.

Thursday Afternoon. Dir. Brian Eno. OPAL Ltd./Sony Corporation of America, 1984.

Timecode. Dir. Mike Figgis. Columbia Pictures, 2000.

Trial and Retribution. ITV1/LaPlante Productions, 1997–2004.

Williams, Raymond. *Television: Technology and Cultural Form.* Hanover and London: Wesleyan University
   Press, 1974.

Woodstock. Dir. Michael Wadleigh. Warner Bros., 1970.

14

Sound in Domestic Virtual Environments

Jeremy Yuille

Artists, filmmakers, game developers, architects, and telephone companies (to name a few) have all at some time wondered how to use sound to help communicate and engage their audiences in the worlds they create. The issues they face and the solutions they reach differ, but all would agree that sound plays an important role in engaging an audience. Just try the standard test of turning the volume down during a horror film to see how important sound is to building a sense of emotional involvement. Films, games, and simulations represent different creative goals and uses of the audiovisual medium with vastly differing production requirements, including a number of off-the-shelf and customized tools. This chapter looks at recent changes in the way artists, designers, and directors use sound in domestic virtual environments, highlighting some current technical and conceptual trends, how these trends link to the past, and where they might take us in the not-too-distant future.

Pioneering artworks such as *Osmose*, Char Davies's 1995 interactive virtual-reality environment installation, originally used separate computers and synthesizers to generate sounds that are "spatially multidimensional," designed to respond to changes in the player's activity within the game in real time. In 1995, this sonic sophistication required individual computers for the sonic and visual elements. In the subsequent decade, this technique of sharing digital signal processing (DSP) over networked processors has become more common, particularly when complex processing or interactivity is required in real time. The fundamental needs of such a system are a high-level schema for communication between computers handling vision and sound and a high-speed network to make this communication possible without noticeable latency. Many works use MIDI for this communication, but more recent applications utilize open sound control (OSC), a protocol developed by Matthew Wright and Adrian Freed for "communication among computers, sound synthesizers, and other multimedia devices that is optimized for modern networking technology."

Works using these techniques often require extended installation time involving specialized technical personnel. Factoring in the cost and logistics of this kind of overhead is not uncommon within the context of a gallery exhibition, particularly for an international show but is not viable in a domestic situation. The limitations imposed by such a separation of audio and visual processing are highlighted if we look at how these kinds of virtual environments have moved into the home in the form of digital games.

In contrast to artists like Davies, who creates innovative works that push the technical possibilities of the time in rarefied gallery environments, the digital games industry relies on a massive user base of standardized equipment being available in the domestic setting. The development of digital games drives the technical innovation, specification, and affordability of game technology add-ons for personal computers. Such add-ons include hardware that offloads processing of sonic and visual media, such as graphics and sound cards, and the associated software that helps applications from games to operating systems use these hardware accelerations.

Games for commercial release usually represent a sizeable investment on the part of the developers in research and development of an "engine" that renders the virtual world of the game. These engines are often built in "middleware," or software that lets a programmer use high-level abstractions instead of having to build a program up from first principles. Middleware makes the development of a complex game less tedious and time-consuming, but it often imposes normative restrictions on audiovisual processing routines. For instance, a programmer might have a great toolbox of ready-made preset routines, but it is quite often impossible to implement anything more exotic or esoteric without rewriting the entire system.

Middleware application programming interfaces (APIs) are designed to accelerate many aspects of game design and experience. APIs range from fully formed macros, or libraries of code that simplify the creation of realistic environments, to methods for communicating with hardware acceleration for procedural effects like reflections, fog, and water or simulation of physical forces like collision, detection, and gravity. The latter is seen to great effect in the Source engine, used to power the game *Half-Life 2* and its subsequent set of related modifications, including *Counter-Strike*. Other important aspects of middleware APIs are the methods they introduce to control how sound behaves in response to gameplay and the virtual environment.

A middleware API will usually have routines or protocols that allow it to "talk" to standardized hardware, like graphics processing units (GPUs) and sound cards. OpenGL is a good example of such a standard protocol. In their user manual for Jitter, a visual addition to Cycling '74's Max/MSP programming environment, J. Bernstein and J. K. Clayton define OpenGL as "a cross-platform standard for drawing 2D and 3D graphics, designed to describe images so that they can be drawn by graphics coprocessors." By affording an abstract, high-level language for describing visual elements in term of their relationship to geometric primitives, OpenGL has dramatically changed the development speed and experiential sophistication of digital games. Games production workflow has adopted the OpenGL paradigm, with modelers, texturers, and lighters working together to produce photo-realistic environments that exploit every technical loophole in order to squeeze an extra frame per second out of a standard GPU.

On the other hand, while sound hardware has become more sophisticated in the past decade, it still lacks the standardized language of primitives that enables the generative power of OpenGL. Couple this limitation in processing power with the commercial reality for games to be played on a wide range of domestic technology, and it is easy to see why sound is usually applied in a simplistic fashion. For instance, most current audio cards (and, subsequently, APIs) support dynamic playback—the looping and mixing of sound files in real time. This means that the game software can use a standard sound card to begin playback of a sound file at any time with dynamic control over the speed (and therefore perceived pitch) of the playback and the level, or loudness, of the sound in relation to other sounds. Many cards and APIs offer advanced methods for spatialization of audio sources, making it possible for individual sounds to appear to originate from different directions in relation to the player. Many standard cards now include reverberation and occlusion algorithms that create approximate acoustic models of the virtual space. For instance, a sound of a conveyor belt in a large, echoing room will move as the player turns and fade dramatically as the player exits the room or walks behind a large object. These methods aid in creating a more "realistic" spatial sound field, which, given that sound is one of our finely developed survival senses, can be of great importance to the success of a digital game.

In most game production environments, sound designers work with a system that can quickly trigger sound and MIDI files, applying real-time effects to create spatial cues in the resulting mix. While certainly a flexible system to compose for, the process of designing for this system is far more linear than the end result because all the source sounds have to be created before the sound design can be actually heard or tested. While this paradigm of playback and collage works well for organizing and structuring sound within a larger design (for instance, music), it does not lend itself well to the creation of complex, sonically realistic soundscapes where the same action can produce subtle variations in sonic response (for instance, the sound of walking in a forest). When looking at the limited toolset available for the game sound designer, it becomes obvious that a high degree of skill, planning, and composition is required to create even the most rudimentary dynamic soundscape.

At this point it is interesting to look at some of the history of computer music, in particular the separation of sound and music and the rise of the personal computer and the electroacoustic composer and performer.

The postwar period in Europe saw two major musical directions evolve, in part driven by the availability of new technologies such as the tape recorder and primitive electronic synthesizers. Both directions strove to move beyond traditional concepts of music and embraced science and production as methods for redefining composition. Simplistically, one movement espoused mathematics, abstraction, and synthesis as its basis, exemplified in works such as Karlheinz Stockhausen's 1959 composition *Kontakte*. The other movement concentrated on the more concrete qualities of sound, researching the phenomenology of listening and perception, well described and presented in the work of Pierre Schaeffer, particularly his 1966 publications *Solfège de l'Objet Sonore* and *Traité des Objets Musicaux*. In these writings, Schaeffer introduces the term "acousmatic" to describe the act of listening to a sound "for its own sake," without reference to what may have created it. This split between abstract and concrete approaches to music is underscored in the way current domestic technology treats the sonic realm. Abstraction is far more easily generalized and accommodated in the playback-and-mixing based engines described earlier. In contrast, the acousmatic qualities of certain sounds are, by their very nature, extremely specific and do not easily lend themselves to the creation of sound engines for dynamic sound design found in games. These concrete approaches have found a home in the more linear sound design fields of film sound design and electroacoustic composition.

Looking at how technical developments have led to the status quo, Curtis Roads's seminal *Computer Music Tutorial* describes early computer music as an elite field characterized by the mainframe computer, research facilities, and composers working with teams of technical assistants. The year 1997 saw two major announcements of personal-computer based DSP tools: MSP, a set of real-time DSP objects for the already popular MAX programming environment, originally created by Miller Puckette, David Zicarelli, and colleagues, and Puckette's PD, a "public domain" cross-platform visual programming environment capable of processing and generating streams of "pure data" in real time. These tools made it possible for composers to rapidly prototype and perform with DSP environments using domestic technology. The model of composer-as-programmer began to branch out into more mainstream musical genres, creating nonlinear hybrids well exemplified by Christian Fennesz's 1997 album *Hotel Paral.lel*, in which sonic structures are constructed from a diverse array of field recordings, data sonification, and digital signal processing. With the introduction of software like MSP and PD, algorithmic approaches to composition and sound design became more widely appreciated and, most importantly, more technically possible and affordable. A fantastic history of this time and its perceived future directions are published in the *Computer Music Journal*'s issue on the 2002 Dartmouth Symposium on the Future of Computer Music Software, particularly Puckette's "Max at Seventeen," and David Zicarelli's "How I Learned to Love a Program That Does Nothing," two reflections on over a decade of the MAX programming environment's effect on computer music.

While environments like MAX/MSP or PD make it relatively easy for the composer to generate anything they can conceive of, each different synthesis method or DSP system has limits on its use in a virtual environment. Methods based on reading a sample or a digitally recorded version of an audio waveform are the most general in their application but have the drawback of needing all the source sounds to pre-exist. You have to make the samples before you can test them out in the sound engine. According to Roads, granular synthesis, a technique of synthesizing sounds from tiny (usually under half a second long) sections or grains of original wave tables, is good for soundscapes that are themselves built from many smaller sources like wind, rain, and insect swarms, but it can also be applied to events like gunshots, opening doors, and engines. More exotic techniques, like Perry Cook's Physically Inspired Stochastic Event Modelling project, in which the qualities of physical surfaces and specific events are abstracted individually, can be used to synthesize a range of specific events like different people's footsteps on a range of surfaces. In *Real Sound Synthesis for Interactive Applications*, Cook gives a thorough account of how such a system could be designed and integrated into virtual environments like games.

As virtual reality and digital games begin to converge in the form of the first-person shooter and massively multiplayer online role-playing games (MMORPG), the need for sophisticated, real-time sonic environments grows. Driven by the market forces of the motion picture industry with the introduction of DVD standards for distribution, multi-channel sound systems have quickly found their place in domestic lounge rooms, and the 5.1 specification of the "home cinema" became the standard for multispeaker spatialization of sound in domestic virtual environments. With the introduction of home computers aimed squarely at the lounge room and game consoles that seamlessly interface with household media libraries and broadband networks, the distinction between computers and entertainment systems is blurring rapidly. Systems that deliver linear content like music and movies sit at a different point on the generic/specific scale than those that deliver interactive content like games and communications. The fact that these systems are rapidly becoming one and the same highlights the need for standardized synthesis models that can accommodate these differing uses with continuous parametric control.

This short history of sound design for domestic virtual environments can be seen as a move toward the development of nonlinear, high-level representations of sound synthesis that can be easily implemented into the abstract, code-based environments of commercial middleware. As toolsets become more "meta," or removed from machine language, and hardware vendors support more standards, domestic sound designs will become more generative and sophisticated in their use of synthesis, not only creating greater powers of persuasion in aural simulation and interactivity but also opening up the possibility that a sound design is a continuous transition between parametric states as opposed to a series of discrete events arranged a priori.

Works Cited

Bernstein, Jeremy, et al. "Jitter Manual." *Cycling '74* 2004 http://www.cycling74.com.

Chion, Michel. *Guide des Objets Sonores*. Paris: Buchat Chastel, 1995.

Chion, Michel., et al. *Audio-vision: Sound on Screen*. New York: Columbia University Press, 1994.

Cook, Perry R. *Real Sound Synthesis for Interactive Applications*. Natick, Mass.: AK Peters, 2000.

Davies, Charlotte. *Osmose*. 1995. http://www.immersence.com/osmose.

Fennesz, Christian. "Hotel Paral.lel." *Mego*, September 1997.

Gorisse, Niels. *CPS, Bonneville*. 2000. http://www.bonnelville.nl/cps.

ISO/IEC. "Overview of the MPEG-4 Standard." ISO/IEC JTC1/SC29/WG11 N3156.

Lyon, Eric. "Dartmouth Symposium on the Future of Computer Music Software: a Panel Discussion." *Computer Music Journal* 26.4 (2000): 13–30.

Paul, Leonard J. "Audio Prototyping with Pure Data." *Gamasutra.com* 10 Nov. 2003 http://www.gamasutra.com/resource_guide/20030528/paul_01.shtml.

Puckette, Miller. "Max at Seventeen." *Computer Music Journal* 26.4 (2002): 31–43.

———. "Pure Data." ICMA International Computer Music Conference, Thessaloniki. 1997.

Puckette, Miller, et al. "MAX/MSP." *Cycling '74* 1997.

Roads, Curtis. *Asynchronous Granular Synthesis: Representations of Musical Signals*. Cambridge, Mass.: MIT Press, 1991.

———. *The Computer Music Tutorial*. Cambridge, Mass.: MIT Press, 1996.

Schaeffer, Pierre. *Solfege de l'objet Sonore*. Paris: INA GRM, 1966.

———. *Traite des objets musicaux*. Paris: INA GRM, 1966.

Stockhausen, Karlheinz. *Kontakte*. Stockhausen-Verlag, 1959.

Wright, Matthew, and Adrian Freed. "Open Sound Control: A New Protocol for Communicating with Sound Synthesizers." ICMA International Computer Music Conference, Thessaloniki. 1997.

Zicarelli, David. "How I Learned to Love a Program That Does Nothing." *Computer Music Journal* 26.4 (2002): 44–51.

15

Getting Real and Feeling in Control: Haptic Interfaces

Joanna Castner Post

Haptics are a collection of tools, including hardware and software, that allows us to feel and manipulate virtual objects. Haptics infuse the digital with the sensual. Imagine experiencing the texture of an ancient manuscript online. Consider the possibilities of adding the tactile to the visual in graphical displays. Speculate about the nature of writing made material in radical new ways. But also think about the social and political implications of embodying our digital creations. What will it mean to make our digital objects and spaces "more real," for example? What is real? If X and Y are real, what does that say about A and B? Haptics are exciting, no doubt, but they also warrant a climate of continuing critique—embodying the digital could involve creating new hierarchies or entrenching old ones.

Let me recommend two particular tools that work in tandem, the HP haptic mouse and the Immersion Desktop 2.1.6 software. They are cheap, accessible tools that can provide many readers with a sense of the possibilities of haptics—in other words, these are tools most readers can actually get their hands on quickly just by running to the local Circuit City.

I'm the kind of person who does not like to buy gadgets I don't need. I bought the HP haptic mouse just for my research on haptics—it was the cheapest and quickest way for me to experience what I'd been reading about. Let me tell you—you must have one. You'll be surprised at the difference. The mouse and software give me a greater sense of control of the Windows interface; even though the development and diffusion of haptics will enable a radical change of such an interface, it makes quite a difference in the status quo as well. This may have something to do with findings such as those of engineer Grigore C. Burdea, who found that haptics could facilitate targeting tasks. For example, users with haptics were able to find and click buttons more quickly because as soon as they felt the feedback as the mouse moved over the button, they were able to click, whereas users without haptics tended to wait till the mouse was directly over the center of the button before clicking.

My Immersion Desktop "theme" is set on "crisp," and the strength is set at 100 percent. I feel and hear a decided "thunk" as I navigate across menus, toolbars, and links. The thunk is lighter and softer when I move over small checkboxes. I feel a slight resistance with a grainy, sandpaper texture when dragging items to the recycle bin. The gallery of example experiences that comes with the software demonstrates textures such as sandpaper, washboard grooves, and the strings of a tennis racket—all are quite realistic. The gallery also allows users to interact haptically with an online poem, to feel the rumble of a digital engine, and to feel a star expanding. Users can choose from a variety of haptic themes, such as "rubbery," "sonic vibe," "metallic," "spongy," and "steel drum." Sonic vibe feels like the twang of a guitar string that is immediately stopped. The rubbery theme feels like a gradual bump with a rather drawn out vibration sound. All the themes come with different sounds, although they are not loud or distracting—they are more like understated sounds you would expect if you actually experienced a concave surface or bump with the mouse. A focus of research in this area is on the interaction among multiple sensory modalities. For example, Marilyn Rose McGee, Philip Gray, and Stephen Brewster of the Glasgow Interactive Systems Group in the Department of Computing Science at the University of Glasgow believe that the combination of audio and haptic information makes the haptic information seem more realistic (par. 10). Based on the mouse and software I use, I would agree, and I would also add that the sound contributes to the sense of feeling in control of the interface—you see the action areas on the screen as usual, but you also feel and hear them.

Haptics and Interfaces

Haptic research is a relatively new and rapidly developing area, so the terminology is not standardized yet. But using Burdea's definitions, there are essentially two kinds of haptic feedback: tactile and force. Tactile feedback involves sensation to the skin, as in all the sensations previously described, and force feedback refers to feelings of resistance, as in feeling the force of a turn of a virtual plane in a flight simulation using a haptic joystick. Haptics are being developed for many different uses: gaming technologies, medical simulations, educational software, and access devices for the blind and visually impaired, for example. Much of the research for these projects is difficult to read for the nonengineer, however. One of the most accessible descriptions of the current applications of haptics comes from the introduction to *Touch in Virtual Environments*, written by the editors of the collection (McLaughlin, Hespanha, and Sukhatme), a group of communication, computer science, and engineering researchers from the Integrated Media Systems Center (IMSC) at the University of Southern California's School of Engineering. I'll list just a few of the examples they describe:

- Haptics are being used in the development of many medical training simulations. Some of the simulations are designed to teach students how to look for and examine liver tumors, perform arthroscopic surgery, cut and suture, examine tooth enamel and dentin, and classify tumors (4). So, to get a sense of what this means, different kinds of virtual liver tumors, for example, are embodied via haptics, and then these virtual objects can be felt and manipulated with haptic devices.
- Haptics are also being used to provide access to virtual, 3D models of museum objects. For example, the Museum of Pure Form is being developed. This museum provides virtual models of sculptures, and visitors can explore them

online and in some cases, via haptics. The editors explain that projects such as this one will also allow museum staff to interact with students online as they explore the virtual objects and to answer questions about those objects and point to their interesting qualities (5). An exhibit of the Museum of Pure Form in 2003 in the Computer Science Department of the University College of London allowed visitors to don virtual reality headpieces and a haptic arm to *feel* works of art (*Museum*).

- Haptics are also transforming online painting, sculpting, and design. For example, a painting application has been developed to allow the visually impaired to paint using different levels of pressure, and haptics have been added to modeling programs to allow engineers to simulate the assembly of mechanical objects (6).

- Haptics are being used to augment scientific visualizations of all kinds as well, allowing researchers to understand and use data in new ways (6).

- Military applications are beginning to profit from haptics. Researchers are working to use haptics to provide information and means of communication when it is difficult to see or hear during battle conditions (7).

- Assistive technologies for the blind and visually impaired have also benefited a great deal from haptics. For example, researchers have developed haptic feedback to allow the blind to read graphical data. Haptics have also been added to other kinds of feedback, such as Braille and speech input, to provide the blind with access to visual interfaces (10).

- Haptics are also facilitating learning. For example, physics professor Miriam Reiner used haptic technology to teach students with no physics background about gravity, friction, elasticity, and inertia. The students used haptic technology to feel the appropriate forces, and then they were instructed to draw diagrams illustrating those forces. Reiner explains that the students were able to draw diagrams that conformed to the correct physics calculations and describe each force without knowing anything about those calculations.

One can interact haptically with a wide range of hardware devices, such as gloves, mice, joysticks, and treadmills. Researchers are developing new hardware and improving current hardware at a rapid rate. PHANTOM—developed by SensAble, a leader in haptics research—is probably the most commonly used haptic output hardware right now for researchers in education and industry. There are several different models available. One of the models, for example, uses a stylus to provide haptic feedback. It seems like almost anything can be designed to provide haptic feedback, however researchers at Purdue's Haptic Interface Research Laboratory, for example, developed one really interesting haptic device. This device is a chair, called the Sensing Chair, that gathers data on pressure points through pressure sensors that act like artificial skin. The chair can process the data in real time. As a result of the real-time capability, various applications are being developed that can make use of this data, such as airbag deployment devices and applications that can help people maintain healthy postures.

Other than the basic description of haptics and their uses, what I would like to highlight here is that haptics provide new ways to interact with computers, digital space, and the virtual objects in digital space. Haptics make it possible to "embody" our virtual objects and digital spaces to better control them, manipulate them, and navigate within them. SensAble explains that 3D software, for example, has advanced rapidly, and its advancements make the mouse and the Windows interface outmoded for work in such environments. They write, "Users are limited by the 25-year old mouse/windows interface which works well for traditional two dimensional tasks like word processing, but it is poorly suited to three dimensional applications. The interface, rather than the machine or the application, constrains professional productivity and creativity" ("Haptic Research" par. 3). They have integrated haptics into their interfaces to allow new ways to control and manipulate 3D objects. The project staff at the MIT Touch Lab, another center of haptics research, writes much the same thing. They explain that the hardware and software enabling haptics has improved a great deal, making entirely new kinds of interactions possible. As a result, haptic interfaces allow people to interact with virtual worlds by enabling them to feel and manipulate virtual objects, and this interaction makes the virtual worlds seem much more real (Durlach et al. par. 5). Karen MacLean, in the Department of Computer Science at the University of British Columbia, is working in the same tradition; she researches and develops alternate ways to interact with computers. Her goal is to develop interactions that don't tie us to a computer desktop, and one way she gets around such interaction is by using haptics. MacLean explains that the current Windows/mouse setup limits the amount of information that can be experienced, the kind of information that can be digitized, and the activities that can be performed using computers due to the pain caused by using the mouse for long periods of time. She writes: "It's more interesting to think of it from the other direction: consider the things we've given up in the physical world which might be nice to have back, but augmented with computation and connectivity. Paintbrush and pencils and musical instruments; a single personal key that lets you into home, car and work and has a distinct feel as you insert it in a lock depending on whether your spouse, a friend or a stranger has been by in your absence; a bank card that feels as heavy as your account balance when you swipe it in the ATM" (par. 2–5).

Thus, haptics enable us to rethink the digital and infuse it with the "real" so we can control it better, use it easier. Ever since Americans moved to the digital environment, some of us have been a bit uncomfortable with its slippery, ephemeral nature. For example, Christina Haas explains that writers, in moving from physical paper to computer screen, say they can't seem to get a sense of the text, which she believes has something to do with the inability to manipulate the physical pages of the text online. Many have wanted to gain the same sense of control over the digital that is *felt* over the physical. The online environment, in comparison with the physical, has seemed disorienting to many, out of their direct control—all that data, after all, is separated from us and our environment and trapped within the computer. Many think of the online environment, then, as a rather confusing, slippery, generally out-of-their-control *place*, a place like a wilderness, or a frontier. Interestingly, we have transferred our metaphorical understanding of nature to cyberspace. Metaphorical understandings allow us to create particular kinds of relationships to nature. Some of these relationships have made it possible to make certain oppressive hierarchies appear to be "natural"; thus there could be some significant social or political consequences of carrying our metaphors and understandings of and relationships with nature to cyberspace.

Metaphorical Understanding and Embodiment

Philosopher Mark Johnson argues that one of the fundamental ways we make meaning is through bodily experience. He writes that as we interact bodily with the world, we create *image schemata*, or organizing structures that help us make meaning out of those interactions. According to Johnson, we then metaphorically project those meanings onto abstract domains to facilitate meaning-making there as well. My argument in this section and the next is that we metaphorically project our understandings of "nature" onto digital spaces and virtual objects to help us understand them.

Johnson explains that we use metaphor to extend our understanding of bodily experience, organized as image schemata, into other, abstract domains. In other words, the way we organize and understand our bodily experience will be used in a metaphorical way to understand abstract concepts. Johnson explains that he uses the term *schemata* differently than it has been commonly used, as "a cluster of knowledge representing a particular generic procedure, object, percept, event, sequence of events, or social situation. This cluster provides a skeleton structure for a concept that can be 'instantiated,' or filled out, with the detailed properties of the particular instance being represented" (M. Johnson 18). For example, Johnson explains that we probably understand many events as fitting into a structured framework. In other words, in certain well-known situations, such as buying a car to use Johnson's example, we expect the characters, such as buyers, managers, car salespeople, and so on, to act in specific ways—buyers and sales people would negotiate the price, the buyer would go on a test-drive (M. Johnson 20).

Johnson, however, wants to extend this definition by focusing on the role image schemata play in organizing experiences arising from interactions between the body and the environment. He explains that this definition is closer to Immanuel Kant's definition of schemata "as structures of imagination that connect concepts with percepts" (21). In other words, Kant used schemata as that which would bridge the gap between the body and its sense perceptions and the mind and its construction of concepts. In contrast to Kant, Johnson wants to get rid of this separation between mind and body by showing that image schemata arise out of bodily experience and work to organize that experience and give it meaning. Then, this experience is metaphorically projected onto abstract domains as a way to make it meaningful as well. In this way there is no separation between mind and body. Instead, as Johnson writes, "our understanding *is* our mode of 'being in the world.' It is the way we are meaningfully situated in our world through our bodily interactions, our cultural institutions, our linguistic tradition, and our historical context" (M. Johnson 102). Understanding, then, is a product of the image schemata as ordering structures and metaphorical projections of those schemata onto all aspects of experience, which he explains are made up of, for example, bodily, perceptual, cultural, linguistic, historical, and economic influences (M. Johnson 105). The metaphorical projections are a means by which we give meaning to our experiences.

Johnson gives the example of our understanding that *more* means *up* as one example of a metaphorical understanding based on the *verticality* image schema. He writes, "The propositional expression 'more is up' is a somewhat misleading shorthand way of naming a complex experiential web of connections that is not itself primarily propositional" (M. Johnson xv). We say, for example, that gas prices keep going up, that work keeps piling up, that inflation is rising, and so on, and Johnson explains that these examples represent a metaphorical understanding that more is up, and this understanding comes from our everyday bodily experiences that are organized into the verticality schema. He writes, "If you add more liquid to a container, the level goes up. If you add more objects to a pile, the level goes up" (M. Johnson xv). Thus, our bodily experiences tell us that more is up, and we metaphorically project that understanding onto abstract concepts such as economics.

According to C. S. Lewis, the roots of our metaphorical understandings of nature as a concept and a thing came from the Greeks, who came to understand nature as a container for everything—nature contains everything and is everything. Neil Evernden explains the significance of this idea: "The possibility of having a *thing* called nature is as significant a development as a fish having a "thing" called water: where there was once an invisible, preconscious medium through which each moved, there is now an object to examine and describe" (20). To illustrate the way the container metaphor works, Johnson gives the following examples: "You wake *out* of a deep sleep [sleep is the container] and peer *out* from beneath the covers [the covers are the container] *into* your room [the room is the container]. You gradually emerge *out* of your stupor, pull yourself *out* from under the covers, climb *into* your robe, stretch *out* your limbs, and walk *in* a daze *out* of the bedroom and *into* the bathroom" (30–31).

A significant outcome of the nature-as-container metaphor was that it allowed us to conceive of particular relationships between humankind and nature. For example, some came to locate us within the container and others outside it. So, in one view, humanity came to be seen as something separate from the nature container, which is a good thing because it was thought humanity should transcend nature, which was viewed as dark, evil, out-of-control, bestial. In another view, nature came to represent the "real." As a result, some believed that humanity fell within this container unavoidably; others believed that humanity could only strive to do so; and still others believed we should create a new container, one filled with a little bit of the "human" and a little bit of the "natural." Evernden illustrates these two general camps, humans as part of nature and humans as separate from nature, by explaining the present-day views of the environmentalists and the industrialists. The environmentalists fall into two basic groups. There are those who believe that humans are a part of nature, so we should live lightly on the land to preserve ourselves. Then there are those who believe that nature represents certain ideals that humans should work toward or work toward incorporating into ourselves, bringing a bit of the "natural" into the human; therefore, we should preserve nature to preserve those ideals as well as the ideal selves we could become. The industrialists, on the other hand, believe that nature is a resource we should use in the service of humankind—humans are apart from and, indeed, more important than nature. These two basic views of nature are in opposition, but those in both camps view nature as a thing, a container—and as a container, we can and do navigate through it, map it, manipulate it, and control it. As I'll show later, these are metaphorical understandings and relationships we take with us to cyberspace.

Remaking Nature into Our Images

The significance of our metaphorical understandings of nature in the present context is two-fold. First, as I argue later, our metaphorical understandings of nature have been transferred to cyberspace. Second, metaphorical understandings of nature have been used to inscribe oppressive hierarchies. As a result of our new ability to infuse the digital (linguistic) with the haptic (material simulation), we have the potential to inscribe oppressive hierarchies in digital media and spread them like never before. Following are examples of the way we have reshaped nature to match our metaphorical understandings of it, instituting and supporting oppressive hierarchies in the process. The examples come from a history by Jennifer Price, beginning with a mid-eighteenth-century school of landscape architecture that determined the design of grounds and lawns in many Western societies but especially in England and America to this day (114). This school of landscape design popularized "natural" grounds and eventually lawns. Such grounds were the embodiment of the container metaphor of nature and the result of particular relationships between people and nature in the late 1700s and early 1800s. "Nature" emerged as the unchanging place amid the many changes occurring in the cities, such as the growth of industry (Price 116). Because nature was thought to be free of humanity and change—timeless, a place apart—it became the authority on everything true and real. As a result, everything "natural" was thought to be "true" and therfore better than the "unnatural," which was considered to be "false" (Price 117).

Creating natural grounds took an enormous amount of time, effort, money, and land. The wealthy, along with the landscape designers, deleted all signs of humanity; they tore down entire villages where tenant farmers lived, chopped down groves of trees and planted others, changed the course of rivers, created lakes, and built new hills (Price 115). Of course, only the wealthy could afford "nature"; the poor lived in the unnatural, "false" cities. This was taken to be a sign that they were not as good as those who could afford to live in nature, and conversely, those living in nature were "naturally" more authoritative, being surrounded by the real and the true—it was as if they were a product of the real and the true, while those in the cities were a product of the unreal and the false (Price 117–18). Price calls this distinction the boundary between "artifice" and "nature"; this is a boundary still in place today, for the most part, although our relationship to nature has changed so that we are able to question boundaries of all kinds.

Controlling Cyberspace by Making It "Real"

This ability to question boundaries brings us back to cyberspace and haptics. Our ability to question the boundaries between the two containers, nature and artifice, as well as our longing to get our hands on and control all things slippery and virtual (making our metaphors and relationship to nature so appropriate and useful), allows us to use our metaphorical understandings of nature, the ultimate "space," to understand cyberspace, even though cyberspace could be described as the ultimate in artifice. As Mark Johnson explains, we often transfer our metaphorical understandings of one thing onto other domains to help us understand them as well. Steven Johnson, in *Interface Culture*, writes about conceptualizing cyberspace as a space. He writes that in 1968, Doug Engelbart, who conceived of computer data as something we might manipulate and created a device with which to manipulate it—the mouse—gave us "our first public glimpse of information-space, and we are still living in its shadow" (11). About that space, he writes, "For the first time, a machine was imagined not as an attachment to our bodies, but as an environment, a space to be explored. You could project yourself into this world, lose your bearings, stumble across things. It was more like a landscape than a machine, a 'city of bits', . . . a place worth living in" (S. Johnson 24). David J. Gunkel writes about the "frontierism rhetorics" "that have invaded and occupied cyberspace from the beginning," referring to terms such as "electronic frontier," "new world," the "cowboys" in William Gibson's *Neuromancer* and the Electronic Frontier Foundation created by John Perry Barlow and Mitchell Kapor (813). This metaphorical transfer has consequences. Gunkel explains, "Consequently, understanding and describing cyberspace through rhetorical devices that are explicitly connected to the age of exploration and the American West opens a discursive and ideological exchange between cyberspace and the hegemony of frontierism" (814).

Haptics represent perhaps the most powerful way to map and manipulate cyberspace and the virtual objects in it since the creation of the mouse and Windows setup. And more than that, haptics allow us to *embody* cyberspace and virtual objects in ways straight out of science fiction. The technology is still evolving, but it's evolving rapidly, as the examples of haptic technology near the beginning of this chapter illustrate. Steven Johnson explains that bitmapping, which refers to the digital location of data in pixels and on screen, "suggested an unlikely alliance of cartography and binary code, an explorer's guide to the new frontier of information" (20). About that digital information, he writes, "Our only access to this parallel universe of zeros and ones runs through the conduit of the computer interface, which means that the most dynamic and innovative region of the modern world reveals itself to us only through the anonymous middlemen of interface design. How we choose to imagine these online communities is obviously a matter of great social and political significance (19)." Indeed.

As illustrated through much of this article, many Americans are working out of a long, Western tradition of thinking about nature in terms of the container metaphor, which makes possible particular relationships with nature. These relationships are dynamic, but one of the significant aspects of creating these relationships is that we tend to manipulate and control nature in ways that seem to make it embody those relationships, as shown in the history of the yard. And as discussed, there were significant social and political implications of that embodiment—those with the money to achieve it were authorized; later, those who understood the embodiment enough to create it in suburbia were also authorized, and all others were transgressors of truth and taste. As a result of recent developments allowing us to rethink the boundaries between artifice and nature, as well as our desire to control our artificial creations as we do the "natural," we have been able to transfer our understanding of nature to cyberspace—one of our last frontiers, to steal from *Star Trek*. How will we embody cyberspace and the virtual objects in it? Who will be authorized and who will be named transgressor? And what will be the social and political implications of these authorizations and namings?

History indicates that our relationships with cyberspace will be in as much flux as those with "real" space, which means that we will embody cyberspace and its virtual objects in many different ways, and again, if history is any indicator, many of these ways will indeed authorize some and exclude others. Professor of political and economic geography Claudio Minca argues that one way we might escape from oppressive metaphysics of representation, with regards to physical space, is by cultivating the attitude that any representation is just one representation out of many of those possible, and one path is just one path among many. So, to relate to cyberspace and the embodiment of it and its virtual objects via haptics, Minca might argue that we should cultivate the attitude that the embodiment we experience in the moment is but one possible embodiment of one possible created relationship to cyberspace out of many possible relationships and embodiments of those relationships.

It is fairly easy to envision this particular proposal in terms of the digital tools mentioned in this chapter—the HP haptic mouse in conjunction with Immersion Desktop 2.1.6. Immersion Desktop signals important parts of the screen by allowing people to feel them with the mouse as they are encountered, but there is no attempt to create a hierarchy of feelings. In other words, there is no sense that a "thunk" is more important than a vibration—all are simply feelings used as signals. And this is all that is necessary for its purpose—to highlight the places within the Windows interface that allow users to *do* something, to pull down a menu, click on a link, or locate a scroll bar. This tool does not attempt to create an order of any kind, but it probably does make people feel more in control of the interface because the action spots are so highlighted. And, as Burdea found in his study, the use of haptics helped people click on action spots more quickly than they did without. As a result, I can see how this tool might make people both feel more in control of digital space and understand that there are multiple ways to interact with this particular interface and none are more desirable than others. So, in this case, I think the design of the tool and its software probably facilitates the kind of attitude and navigation arising from Minca's ideas.

But his proposal doesn't really work for virtual objects embodied with the use of haptics—3D objects we can manipulate and touch. For this more complicated issue, Kristie S. Fleckenstein's "biorhetoric" has fascinating implications. Fleckenstein proposes biorhetoric as a way to bring about transformation. She explains that biorhetoric comes from that space in which materiality and semiosis blur. Going back to Mark Johnson's ideas, the way we speak and think about the material world is a result of our bodily interactions with it. As a result, the two elements, the semiotic and material, are necessarily intertwined in complex ways in our lives. Fleckenstein argues that the semiotic-material nexus might be used in a transformative way because, as we move back and forth between our namings and understandings of materiality, we can use metaphor to imagine new ways of thinking and being in the world.

The incorporation of haptics into linguistic cyberspace offers people a new way to examine closely the systems of meaning we have created out of the material-semiotic dialectic and then ideally to work to transform those systems of meaning. Choosing which aspects of our digital creations to highlight with haptic feeling is a sort of naming. Such naming must be drawn from the material, and in this case, the kinds of namings we can create will be constrained by both technology and programming. But haptic namings must also be based in the "normal" interplay between semiosis and materiality. In other words, what we choose to highlight with haptics and how we choose to make it feel will be the result of our experience apart from technology and programming as well as the nature of the technology and programming. Thus, haptics seem to call for, or allow, an unusual examination of the metaphors we use to understand those parts of materiality we have named and an opportunity and new technological ability to imagine new metaphors, new understandings of materiality and reality. Consider these hypothetical scenarios:

- The military incorporates haptic technology into their urban warfare training simulations. To adequately prepare soldiers for the battlefield, the hand-to-hand combat techniques used to kill enemies must feel real. Soldiers in the simulation feel the blade entering the flesh, the shock of blade striking bone. Will these kinds of simulations make war more or less desirable? Will they desensitize soldiers to killing? Or will they highlight the horror of war and act as a deterrent? Would it be possible to operate in the spaces between semiosis and materiality to create transformative ideas of war through such simulations or through non-military countersimulations?

- The public school system uses digital frogs instead of real ones for student dissecting practice. Haptics allow students to feel the knife entering frog flesh, to cut it, and then to move organs aside to see other organs. Students may dissect over and over again, as the digital frogs represent unlimited resources. Will such a practice make nature seem more or less precious? Is there a way to construct such simulations to create transformative notions of nature?

- Medical schools make use of haptic technology in their training simulations to teach how to search for liver tumors. Haptics allow students to feel flesh and tumors or other growths. Would working on virtual people make the bodies of living patients more or less real? Would such work make doctors more or less compassionate? And could such simulations be made to serve transformative purposes?

Such issues seem important ones to consider in an era that is beginning to embody its virtual objects as well as spaces, which are conceptualized as "real spaces," spaces we remake to embody our understandings and relationships to nature, our understandings and relationships that have historically authorized some and excluded others. What will be embodied? What will it mean to make a virtual object "more real" with haptics? And what will our choices signify? If history repeats itself, it seems almost certain that answers to these questions will ultimately reveal new kinds of hierarchies and exclusions. I do think the current is climate allowing people to question the nature of boundaries, proposals such as Minca's, and theories such as Fleckenstein's biorhetoric offer hope for avoiding some oppression. But I also think vigilance and continuing critique will be necessary for making the newly embodied cyberspace a place worth living in.

Works Cited

Burdea, Grigore C. *Force and Touch Feedback for Virtual Reality.* New York: John Wiley and Sons, 1996.

Durlach, N. I., et al. "Virtual Environment Technology for Training." *MIT Touch Lab.* 2002. 15 Oct. 2002 http://touchlab.mit.edu/oldresearch/currentwork/applications/VETT/index.html.

Evernden, Neil. *The Social Creation of Nature.* Baltimore: Johns Hopkins University Press, 1992.

Fleckenstein, Kristie S. "Bodysigns: A Biorhetoric for Change." *Journal of Advanced Composition* 21.4 (2001): 761–90.

Gunkel, David J. "Hacking Cyberspace." *Journal of Advanced Composition* 20.4 (2000): 797–823.

Haas, Christina. *Writing Technology: Studies on the Materiality of Literacy.* Mahwah, N.J.: Lawrence Erlbaum, 1996.

"Haptic Research." *SensAble Technologies.* 15 Oct. 2002 http://www.sensable.com/haptics/haptics.html.

Johnson, Mark. *The Body in the Mind.* Chicago: University of Chicago Press, 1987.

Johnson, Steven. *Interface Culture*. New York: Basic, 1997.

Lewis, C. S. *Studies in Words*. 2nd ed. Cambridge: Cambridge University Press, 1967.

MacLean, Karon. "Research: Physical User Interfaces." MacLean Home Page, Department of Computer Science, University of British Columbia. 2002. 15 Oct. 2002 http://www.cs.ubc.ca/spider/maclean/research/research.html.

McGee, Marilyn Rose, Philip Gray, and Stephen Brewster. "Feeling Rough: Multimodal Perception of Virtual Roughness." 2001 Eurohaptics Conference. 15 Oct. 2002 http://www.dcs.gla.ac.uk/~mcgeemr/euro-haptics-final.doc.

McLaughlin, Margaret L., João P. Hespanha, and Gaurav S. Sukhatme. *Touch in Virtual Environments*. Upper Saddle River, N.J.: Prentice Hall, 2002.

Minca, Claudio. "Postmodern Temptations." *Postmodern Geography Theory and Praxis*. Ed. Claudio Minca. Malden, Mass.: Blackwell, 2001. 196–225.

The Museum of Pure Form. 2004. Scuola Superiore di Studi Universitari e di Perfezionamento. 13 Aug. 2007 http://www.pureform.org/project.htm.

"PHANTOM." *SensAble Technologies*. 15 Oct. 2002. http://www.sensable.com/haptics/products/phantom.html.

Price, Jennifer. *Flight Maps: Adventures with Nature in Modern America*. New York: Basic, 1999.

Reiner, Miriam. "Conceptual Construction of Fields Through Tactile Interface." *Interactive Learning Environments* 7.1 (2000): 31–55.

"Sensing Chair." *Haptic Interface Research Laboratory. Purdue University*. 2001. 15 Oct. 2002 http://www.ecn.purdue.edu/HIRL/projects_chair.html.

16

Digital Craft and Digital Touch:
Hands-on Design with an "Undo" Button

Mark Paterson

In terms of artistic applications, craft, and industrial design, the implications for the convergence of digital craft with digital touch are significant. As digital designer Malcolm McCullough argues, "Increasingly computing shows promise of becoming the medium that could reunite visual thinking with manual dexterity and practiced knowledge" (50). The technologies of touch, or "haptics," are a rapidly developing area where such manual dexterity and practiced knowledge, so necessary for the kinds of direct contact and manipulation of materials that characterize craft and hands-on design, can enter the digital realm. In past decades haptic technologies have been quietly proliferating, finding uses in such diverse areas as surgical and military training, long-distance keyhole surgery, mine clearance, Internet sex, undersea and interplanetary exploration, and video games (Hannaford; Stone; Arthur). Haptic technologies are literally making their presence felt everywhere. We are familiar with elementary force feedback devices for video game consoles, which provide basic haptic feedback to enhance the sense of immersion within the game world. Elsewhere, haptics are being integrated into the computer interface on our desktops, through mice and joysticks to more complicated design tools, and are even transforming our experience of virtual environments. This chapter follows one particular haptic device, the PHANTOM Desktop haptic interface, through a range of different applications and design contexts and considers the implications for the convergence of digital craft with digital touch through such emerging haptic devices.

As an addition to the sense of touch, haptics allow, in Steven Johnson's words, a sense of the "direct manipulation" of objects, where "the user makes things happen in an immediate . . . way" (179) and therefore a sense of being immersed, of being engaged in the task at hand. Concentrating on the PHANTOM, we shall see how it allows a sense of presence and of the direct manipulation of the material, which are undoubtedly desirable qualities for digital artisans pursuing their craft. The marrying of a visual display with a haptic interface is far more intuitive for the feeling of direct manipulation and shaping of virtual forms. Thus, the sense of direct manipulation and crafting of material that characterizes traditional craft and design techniques can be carried forward into the digital realm through haptics, with the ability to mold "virtual clay" as exemplified by one application. With these haptic devices, there are implications for wholly new virtual tools as well as faithful reproduction of sensations of texture and the properties of the materials.

After a section briefly describing the PHANTOM device and its haptic capabilities, we then focus on two contexts where haptics are altering design and production, both utilizing the PHANTOM. Firstly, the Tacitus Project in Edinburgh is conducting research on the human-computer interface and how haptics can enhance small-scale digital crafts and enhance the process of hands-on design within a digital medium. Secondly, in the larger-scale industrial setting of digital craft and design for British ceramics company Wedgwood, haptics allow the manipulation and shaping of "virtual clay." This "virtual sculpting" through haptic digital design is virtualizing and revitalizing sectors of the ceramics industry. It allows a form of "virtual prototyping" in the context of industrial design, meaning that innovative designs can be molded and tried, shortening the production cycle. Such significant advantages have been noticed by the automotive and footwear industries, which have also been using these haptic interfaces.

The PHANTOM: Hands-on Design

Deriving from the Greek word *haptesthai*, meaning "of, pertaining to, or relating to the sense of touch or tactile sensations," haptics is the study of touch and how we interact with the world through this sensory mode ("Haptics"). From vibrating mobile phones to video game controllers, haptic technologies are now seemingly ubiquitous. Whereas the current generation of video game console controllers offer basic haptic functionality by vibrating in response to some actions in the game, more sophisticated haptic devices such as the PHANTOM (see Figure 1), invented by Thomas Massie and Kennth Salisbury in 1993 at the Massachusetts Institute of Technolgy (MIT), allow a far higher fidelity of tactile sensation, including the simulation of 3D shapes, textures, and even virtual clay (Amato; Sener, Wormald, Campbell). These haptic interfaces, used for computer-aided design and manufacture (CAD/CAM), are becoming increasingly available as their cost decreases. PHANTOM consists of a stylus attached to a fully moveable arm, and a three-dimensional area of free movement on the desktop is permitted. A haptic object is modeled as a set of resistances to hand movements, and these resistances are induced by electric motors in the arm (i.e., force feedback). In addition, through carefully controlled vibration, the simulation of textures on the object's surface can occur. In this way, touching a virtual object becomes startlingly lifelike. Their mimetic quality, the modeling of real-world tactile properties, is only one aspect of these devices, however. Their use in industrial design, artistic applications, and museum interaction entails some innovative uses for touch in a craft and design context.

Traditional CAD/CAM utilizes a predominantly visual interface, requiring movement of the mouse on a two-dimensional (2D) desktop to be translated into a three-dimensional (3D) virtual model within a visual display. Traditional input through the keyboard and mouse, as Robert Stone argues, is regarded as "very user-unfriendly and completely inadequate for creating organic, free-form shapes" ("MUSE Virtual Presence"). Haptic interfaces offer an alternative to these methods of input, representing

Figure 1.   The PHANTOM at Certec, Lund, Sweden (Photograph by the author; Mark Paterson, copyright 2000.)

a return to more traditional hands-on craft methods, only within a digital medium. This chapter is concerned with how design and craft practices, that is, manual dexterity and practiced knowledge, are being transferred into a digital medium through the introduction of a particular 3D haptic interface, the PHANTOM. The use of any such haptic interface within 3D modeling software effectively provides hands-on design with an "undo" button. It gives us the ability to play with the virtual tactile properties of an artistic medium, such as sculpting with virtual clay. The potential of 3D haptic interfaces allows both possible and impossible physics, a range of new virtual tools, and "the creation of entirely new ways of using touch" (Mullins 38).

But before we consider the new, exciting possibilities of digital touch, we must ask: what is the relation between touch in traditional craft and digital touch? Until a few years ago, few would have seen any continuity or connection. For example, in thinking about the difference between traditional craftwork and the use of digital technology, McCullough asks: "What good are computers, except perhaps for mundane documentation, if you cannot even touch your work? The fact that traditional craft endures at all is because it satisfies some deep need for direct experience—and most computers are not yet providing that experience" (25). Yet we can document some technologies that go some way toward this. As the fidelity of haptic devices increases, the content of digital sensation alters such that the direct feeling of the weighty material, so necessary for intricate craftwork with the hands, can be reproduced. In other words, there are continuities between traditional handcraft—or hands-on design—and recent developments in haptic technologies. There are also *dis*continuities that provide significant advantages over traditional methods, as will become apparent.

Tacitus: Tactile Technologies and Tacit Knowledge

The Tacitus project, which ran from 2001–2004, was a combined venture of the Edinburgh College of Art and the Edinburgh Virtual Environment Centre (EdVEC) and was concerned with the role of haptic technologies in digital design and the applied arts. Sharing McCullough's concern with manual dexterity and practiced knowledge, this research project specifically addressed the role that innovative haptic technologies could take in the craft and design process. The project addressed how 3D haptic and multisensory computer content is capable of meeting the needs of applied artists, craftspeople, and designers and hence of aiding the creative processes of art and design.

A central consideration was the role of multisensory experiences in the shaping and manipulation of materials, thereby making such tacit knowledge concerning this process more explicit in order to transfer or remediate it within the digital realm: "An applied artist's instinctive grasp of constructing and visualising in three dimensions, their spatial thinking and sense of touch are integral to their process of creativity. Makers combine all their sensory modalities, such as sight, hand motions, and sound in order to explore and bring intended qualities to the object they are making" (Shillito et al., "Tacitus Project," *Digital Creativity* 196). To take the creative working practices of craftspeople into the computer desktop would be the ultimate test of a haptic technology. To those craftspeople wary of computer interfaces, the project asks how they can "bring their awareness of space, mass and form to the virtual environment" (Shillito et al. "Tacitus Project," *Digital Creativity* 197) to a desktop-based haptic interface to recreate a sense of the material, of ponderability, which is especially familiar to sculptors.

Using the PHANTOM device in conjunction with a visual display directly superimposed over it means there is direct collocation of the visual and the haptic. This arrangement of haptic device and superimposed screen is a commercial product known as the Reachin Display from the Swedish company Reachin Technologies (see Figure 2), marketed as a solution for industrial design and medical simulation. According to Reachin, "The natural computer interface realized through Reachin Display makes the user feel, see and interact with objects just like in the real world. . . . The innovative use of a semi-transparent mirror creates an interface where graphics and haptics are co-located—the user can see and feel the object in the same place" ("Reachin Display"). To manipulate a virtual object with the hand and have the image change accordingly on the screen through the haptic device produces a credible sensation of a tangible virtual object. A medical simulation of injecting a hand with a hypodermic needle that I tried at Reachin's Stockholm headquarters in 2000 provided enough visual detail to see the veins on a 3D model of a hand, along with enough haptic detail that the skin's springiness and elasticity was felt through the PHANTOM underneath. When a needle was inserted into the skin, the puncturing moment was reproduced through the force feedback in the PHANTOM below. These sensations were felt not simply as a localized kinaesthetic force by the fingertips alone but as a visceral feeling, that is, the synthesis of the visual and the haptic mimicking the sensation of injecting into the springiness and resistance of flesh.

In the context of virtual craft, therefore, such properties as elasticity and weightiness of materials can similarly be modeled. In an interview conducted at their headquarters, one of the Reachin team expressed this crossover between digital craft and industrial design, exposing the frustrations of working with traditional CAD/CAM and explaining how this can be resolved through haptic digital tools. According to Tomer Shalit, the products manager at Reachin, industrial designers who tried out the Reachin Display had "the ability to work with digital tools but still be creative." Despite the company's adherence to realism, he went on to say, "When you see our vision, it's [like] having a creative tool" (Tomer [English corrected]). So far this sounds almost conventional, the remediation of traditional tool use into a virtual environment. Yet, just like other forms of remediation, the discontinuities are as fascinating and important as the continuities. Firstly, I

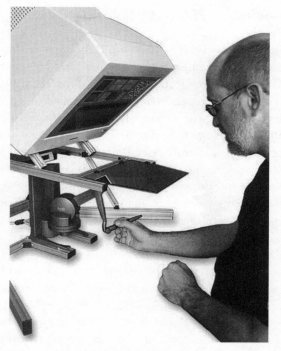

Figure 2.   The Reachin Display, a combined visual and haptic desktop (Copyright 2006, Reachin Technologies.)

will consider the continuities and then the discontinuities and their implications for digital design.

Continuities and Discontinuities

The ability to adopt the traditional hand postures and tool use associated with craftspeople and designers and to translate this into the digital realm is one way that the continuity between traditional techniques and the digital medium has been achieved. But in the craft context, the posture of sitting at a desktop and manipulating a haptic display of a virtual object with the hands while being able to look down upon a visual display of the virtual object with the eyes is an effective enough approximation of the traditional posture and attitude (Shillito et al., "Tacitus Project," Eurohaptics 3). The Reachin Display is designed for the simulation of close proximity handwork, whether of a surgical, medical, military, or engineering kind.

To understand the continuities between traditional tool use and this haptic desktop is to work with and understand material constraints and simulate them effectively. Through combined sensory effects and especially visual-haptic collocation, a sense of interacting and working with the material is possible, and in turn, this can encourage the necessary multisensory skills to achieve the creative intent. For example, a silversmith must use audible, tactile, and kinaesthetic cues in shaping the material. McCullough suggests that the sense of bending or morphing materials with the free-form gestures of sculpture requires that feeling of ponderability, of "palpable mass" (102). Through combined sensory effects, haptic technologies such as the PHANTOM and the Reachin Display have shown some measure of success in this.

Merely imitating traditional craft and design techniques or simply remediating them in digital form can only go so far. Despite the remarkably lifelike sensations that can be achieved through visual-haptic collocation, the goal of the Tacitus project was to create a fluid and intuitive virtual environment that could be applied to a variety of creative disciplines, a multisensory environment within which craftspersons or designers could bring their experience and knowledge to extend their levels of creativity within a digital context (Shillito et al. "Tacitus Project," Eurohaptics 4). In this way, some of the multifarious, tacit knowledge involved in working directly with materials could then be applied to the multisensory virtual environment. But rather than simply mimicking the postures and tools of creative practice, the Tacitus project sought to positively enhance those practices and processes, bringing some advantages of the digital medium to them. For example, based on extended observation of craftspeople and follow-up interviews, the Tacitus team wished to develop "virtual 'hand tools' to enhance the creative practice of applied artists" (Shillito et al. "Tacitus Project," *Digital Creativity* 201). The possibility of creating new ways of working with materials, of blending traditional approaches with innovative approaches and novel tools only possible in a digital medium, highlights these continuities and discontinuities with traditional craft. Virtual tools may simulate traditional tools for ease of approach and fluidity of movement, but they can depart from that realism to yield advantages associated with the digital medium: "[S]ay you're doing sculpting, you want something that's roughly like clay so you can understand how to manipulate it, particularly if you're aiming it at someone who works with clay. There's an interesting trade-off between . . . where to go for realism and where to go away from realism in order to give more flexibility . . . than you can have in the real world. The undo button is a simple example of that" (McLaughlin).

One significant advantage of haptic digital tools is that there is still hands-on design, only with an "undo" button. There are other advantages, such as the possibility of novel virtual tools and the alteration of the properties of the virtual material, which brings us neatly to a more detailed consideration of "virtual clay" and how a creative means of input may influence industrial design and production.

Virtual Clay: Haptic Tools and Industrial Design

While Edinburgh's Tacitus project has taken an experimental approach to using haptics for design, the PHANTOM is also involved in more industrial design settings, for example in virtualizing and revitalizing the British ceramics industry. The Midlands is an area historically identified with the pottery and ceramics industry and, like many other examples of industry in the United Kingdom, has been in decline. Nevertheless, the prestigious ceramics company Wedgwood remains, and there is evidence of revitalization of this industry through the use of haptic devices with virtual clay modeling, detailed by Stone. The design of ceramics at Hothouse Centre for Ceramics Design in Stoke, which contracts to Wedgwood, has been given a new lease on life through digital design techniques, using the PHANTOM desktop haptic interface (see also Paterson, "Digital Touch" and "Feel the Presence").

As with the Tacitus project, visual-haptic collocation—the marrying of visual display and a haptic interface—is far more intuitive for design and craft, for the feeling of direct experience with the manipulation and shaping of forms. This visual-haptic collocation system at Hothouse simulates the feel of the on-screen 3D surfaces, so that "potters and sculptors can 'touch' and 'feel' the models they are creating," according to MuseTech ("MUSE Virtual Presence"). This system is based on commercially available 3D-design software called FreeForm, specifically designed for the PHANTOM. The marketing of FreeForm to 3D CAD/CAM sectors creates much of the hitherto missing sense of touch in the manipulation and crafting of objects in previous CAD/CAM products and echoes the free-form gestures that sculptors and craftspeople depend upon. The software directly mimics the techniques traditionally used in clay modeling. As Sener and colleagues describe, "An imported model or a user-defined block of 'clay' can be modified using different tools such as a ball, square block and scrape via the PHANTOM" (548).

Hothouse is therefore marrying the small-scale, hands-on process of digital craft, as at Tacitus, with the larger processes of industrial design and manufacture. Models can be created through modeling in "virtual clay" and then output to manufacturing tools as part of an automated process. Those more "organic" shapes can be created and literally digitally manipulated through the software, and this impacts the way that digital craft proceeds. But for the process of digital craft to be truly digital, the necessary aspect of direct experience that McCullough identified, actual manipulation, must involve the hands as well as the eyes. Using a haptic device collapses the distance between the virtual object and its representation on the screen so that it becomes directly manipulable through the haptic interface software. According to manufacturers SensAble Technologies, what underlies the FreeForm software is 3D Touch, a basic "physics of touch" engine created by a set of algorithms that generate the force feedback through servomotor interactions. It is this layer of software that gives the illusion of solidity, elasticity, texture, and the manifold tactile properties of virtual objects in conjunction with the PHANTOM hardware. John Ranta of SensAble explains this more intuitive method of control: "Designers want to be able to reach in and push down on a surface and have a contour change—just the way they would with a piece of clay—and feel the surface, the spring force, and the resistance, and at the same time watch the surface update its shape" (qtd. in Mahoney 42).

Feeling the force and resistance of tools and materials, a necessary component of the crafting experience, is accomplished, therefore, through force feedback, the feeling of resistance on the haptic device. With such 3D force feedback, artists can use the handheld stylus of the PHANTOM like a brush, applying virtual paint onto 3D objects, so that forces felt through the stylus give enough tactile feedback to make the virtual brush flare or feather with slight changes in pressure, just as actual paintbrushes do. As Hodges explains, "The user can rotate this virtual brush freely, sensing surface textures as well as the thickness of the virtual paint" (50). These haptic crafting tools are applicable to computer graphic animation, where, for example, conceptual models for the film *Chicken Run* were manipulated by Aardman Animations using the PHANTOM (Kay).

Additionally, the software allows the creation of wholly new artistic tools to incorporate into the design process, a possibility raised previously. The tension between the desire for verisimilitude in the simulation of touch and the simultaneous desire to utilize the advantages of the digital medium seems to characterize a central feature of haptic digital tools, whether for small-scale sculpting or for large-scale industrial design. Apart from the ability to invent wholly new digital tools to work with, another advantage of the digital medium is the ability to dynamically alter the properties of the material being crafted on the fly. This entails a different orientation toward materials and form, since certain shapes and levels of detail are based on the feedback of particular materials. If virtual clay can be made more or less malleable within the design software through the movement of a digital sliding button (as in Photoshop), then the artisan's choice of appropriate materials in order to create the intended form, shape, and level of detail of an object is not so crucial. Whether for a sculptor or an industrial designer, the constraints of the material being worked would entail different sensory feedback, as we saw with Shillito and colleagues' remarks on the sensory modalities and creativity. By altering the properties of the virtual clay, different sensory feedback ensues. The difficulties of working within particular material constraints disappear and new imaginative possibilities start to emerge. As Tomer of Reachin puts it, "so the idea of being able to sculpt something and then decide, 'no I want this object, I want it to be a different material,' to feedback to me in a different way, that's very exciting and then you can start, of course, start getting entirely different materials. I mean you could sculpt in floating air" (Tomer).

Hands-on Design—with an "Undo" Button

From the specific cases of digital craft and design, I wish to conclude by asking two more general questions: How does haptics affect cultural contexts? And how will haptics transform future relationships to cultural environments?

Firstly, there are implications for the user interface in general. What McCullough required from a computer was the sense of direct manipulation of the material, even if in a virtual form. Likewise, the importance of haptic devices lies not simply in its specialized use by the digital designers and artisans discussed; it also will alter the human-computer interface (HCI) in other more general applications. In the same way that the mouse shifted the computer desktop into a graphical user interface (GUI) with windows and pointing and clicking at Xerox Palo Alto Research Center (PARC), perhaps we will see the onset of the tangible user interface (TUI). At least, this is the thought of Hiroshi Ishii and Brygg Ullmer (see also Ullmer and Ishii, "Emerging Frameworks") at the MIT Media Lab. The requirement for "direct manipulation" will be felt not only by design professionals, as we have examined, but also may filter through to the general user.

With the convergence of all kinds of digital media and the problems of accessing this endlessly proliferating information space, the usual 2D representations of data as, for example, windows and folders are fast becoming inadequate. The sheer volume of data and the increasingly vast numbers of files and their larger size requires that access be achieved otherwise. Whether in an abstract form, as 3D spatial representations, or through visual-haptic collocation with a haptic device, such as a stylus or a dataglove, the sense of direct manipulation of data has implications for the human-computer interface in general. Steven Spielberg's 2002 science-fiction film *Minority Report* addresses this directly, based on consultations with the MIT Media Lab. Data is streamed and literally manipulated (from the Latin *manus*, meaning "hand") with a very unusual interface, using gloves to spatially organize the dataflow on the screen. More generally, this is indicative of a trend in human-computer interface design where, in Johnson's words, "we can be sure that the exploratory, spatial quality of the medium—the *haptics* of information-space—will be of enormous importance" (221; original emphasis). In other words, we can speak of the "look" and "feel" of the user interface both metaphorically and literally.

Secondly, another startling possibility emerges: the ability for others to share the "feel" of an object over a distributed network. Communicating the feel of objects, the sense of their presence in a distributed network, among several computers and users is of great interest (Paterson, *Environment and Planning*). This mutual touch, where one user can feel another, is ideal not just for collaborative design. It could also be used for training purposes, where haptic experiences could be recorded and played back. A new master-apprentice relationship that includes the feel of the materials and the feel of the teacher's hand could result. Ivan Amato thinks it would allow "in-the-hands knowledge of artists and artisans to become digitised and preserved for future students" (71). Thus, product design, craftwork, animation, and a whole host of creative applications, as well as the teaching of these, can be enhanced by the haptic interface.

If the origin of haptics and force feedback lies in bringing distant objects to life, however, then another aspect of haptics is the intimacy of touch, the prehensile and exploratory space of touch that is within reach. When using a visual-haptic technology, the potter feels no separation between himself or herself, the virtual clay, and the task in hand. In this regard, the efficiency of a computer interface can be measured by the ability to feel an object and to immerse oneself in the task as one would in the real world. The crucial difference is that, in the digital world, there is an "undo" button.

Notes

1.   The acronym PHANTOM stands for Personal HAptic iNTerface Mechanism. See Massie and Salisbury; Salisbury.

2.   According to SensAble and Reachin, large companies such as Nike and Saab-Scania employ these haptic interfaces. See http://www.sensable.com and http://www.reachin.se (last accessed 30 June 2005).

3.   Although initially their cost was prohibitive, around US$19,000 in 2001, a recent variant, the Omni, is now available for £1,470 (US$2,600) (e-mail correspondence with David Hendon of Virtalis, the UK distributor for SensAble, 25 July 2005).

4.   The project's Web site is http://www.eca.ac.uk/tacitus/index.htm (last accessed 22 June 2005).

5.   John Underkoffler, consultant to Spielberg on the film, worked for the MIT Media Lab. A comparable system made by Raytheon (also Boston-based) is currently under development using gesture tracking, according to *New Scientist* (Knight).

Works Cited

Amato, Ivan. "Touchy Subjects: From Digital Clay to the 'nanoManipulator.'" *Technology Review* Apr. 2001: 70–71.

Arthur, Charles. "Touching Moment 3,000 Miles Apart Becomes a Virtual Reality." *The Independent* 30 Oct. 2002: 7.

Hannaford, Blake. "Feeling is Believing: A History of Telerobotics." *The Robot in the Garden: Telerobotics and Telepistemology in the Age of the Internet.* Ed. K. Goldberg. London: MIT Press, 2000. 246–75.

"Haptics." Oxford English Dictionary. 2nd ed. 1989.

Hodges, Mark. "It Just Feels Right." *Computer Graphics World* 21.10 (1998): 48–56.

Ishii, Hiroshi, and Brygg Ullmer. "Tangible Bits: Towards Seamless Interfaces between People, Bits, and Atoms." *CHI '97 Conference Proceedings: Human Factors in Computing Systems*. Reading, Mass.: Addison-Wesley, 1997. 234–41.

Johnson, Steven. *Interface Culture: How New Technology Transforms the Way We Create and Communicate*. New York: Basic Books, 1997.

Kay, Russell. "The Sensual Computer: High Touch, High-Tech." *Computer World*. 9 Oct. 2000. 16 Feb 2004 http://www.computerworld.com/industrytopics/healthcare/story/0,10801,52067,00.html.

Knight, Will. "*Minority Report* Interface Created for US Military." *New Scientist* 15 April 2005. 30 July 2005 http://www.newscientist.com/article.ns?id=dn7271.

Mahoney, Diana Phillips. "The Power of Touch." *Computer Graphics World* 20.8 (1997): 41–48.

Massie, Thomas H., and J. K. Salisbury. "The PHANToM™ Haptic Interface: A Device for Probing Virtual Objects." *Proceedings of the ASME Synposium on Haptic Interfaces for Virtual Environment and Teleoperator Systems*. Chicago: ASME, 1994. 295–302.

McCullough, Malcolm. *Abstracting Craft: The Practiced Digital Hand*. London: MIT Press, 1998.

McLaughlin, John. Personal interview. 13 September 2000.

Mullins, Justin. "Hear me, see me, touch me." *New Scientist* 7 (Nov. 1999): 36–39.

"MUSE Virtual Presence Brings the Touch and Feel of Clay to 3D Ceramics Design." *PR Newswire* 23 Aug. 2000. 27 Jan. 2001 http://www.musetech.com/html/press/ceramics.html.

Paterson, Mark. "Digital Touch." *The Book of Touch*. Ed. Constance Classen. Sensory Formations. Oxford: Berg, 2005. 431–36.

———. "Feel the Presence: Technologies of Touch and Distance." *Environment and Planning D: Society and Space* 24.5 (2006): 691–708.

"Reachin Display: For Co-located Graphics and Haptics." *Reachin—Products*. 2006. 6 Feb. 2006 http://www.reachin.se/products/Reachindisplay/.

Salisbury, J. K. "Haptics: The Technology of Touch." *SensAble Technologies*. 1995. 17 Feb 2004 http://www.sensable.com/products/datafiles/phantom_ghost/Salisbury_Haptics95.pdf.

Sener, Bahar, Paul W. Wormald, and Ian Campbell. "Towards 'Virtual Clay' Modeling: Challenges and Recommendations: A Brief Summary of the Literature." *Proceedings of the DESIGN 2002 7th International Design Conference*. Dubrovnik: Design Society, 2002. 545–50.

"SensAble." *SensAble Technologies*. 14 May 2001 http://www.sensable.com.

Shalit, Tomer. Personal interview. 13 September 2000.

Shillito, Ann Marie, Karin Paynter, Steven Wall, and Mark Wright. "Tacitus Project: Identifying multisensory perceptions in creative 3D practice for the development of a haptic computing system for applied artists." *Digital Creativity Journal* 12.5 (2001): 195–203.

———. "Tacitus Project: Identifying multisensory perceptions in creative 3D practice for the development of a haptic computing system for applied artists." Eurohaptics 2001. Birmingham, UK. 1–4 July 2001. 22 June 2005. http://www.eurohaptics.vision.ee.ethz.ch/2001/shillito.pdf.

Stone, Robert. "Haptic Feedback: A Potted History From Telepresence to Virtual Reality." *Proceedings of First International Workshop on Haptic Human-Computer Interaction*. Eds. S. Brewster and R. Murray-Smith. Berlin: Springer-Verlag, 2001. 1–16.

Ullmer, Brygg, and Hiroshi Ishii. "Emerging Frameworks for Tangible User Interfaces." *Human-Computer Interaction in the New Millenium*. Ed. J. M. Carroll. Reading, Mass.: Addison-Wesley, 2001. 579–601.

Contributors

Wendy Warren Austin is associate professor at Edinboro University of Pennsylvania, where she teaches technical writing and composition. She has published articles in *Kairos: A Journal of Rhetoric, Technology, and Pedagogy, Composition Studies,* and *Intercom,* and in the edited collection *Internet-Based Workplace Communications: Industry and Academic Applications,* edited by Kirk St. Amant and Pavel Zemliansky.

Jim Bizzocchi is assistant professor at the School of Interactive Arts and Technology, Simon Fraser University. He teaches courses in new media narrative, game design, and video production.

Collin Gifford Brooke is associate professor of rhetoric and writing at Syracuse University, where he directs the Composition and Cultural Rhetoric doctoral program. His first book, *Lingua Fracta: Towards a Rhetoric of New Media* is forthcoming.

Paul Cesarini is assistant professor in the Department of Visual Communication and Technology Education at Bowling Green State University. He has written for a wide variety of academic journals and trade publications, including *The Chronicle of Higher Education, The Journal on Excellence in College Teaching, EDUCAUSE Review, ACADEME, The IT Manager's Journal, Computers and Composition Online, The National Association of Industrial Technology, Journal of Industrial Technology, Journal of Literacy and Technology, International Society for Exploring Teaching and Learning,* and *The Review.*

Veronique Chance is a media artist completing a PhD at Goldsmiths College, University of London, where she has also taught in the Fine Art and History of Art Joint Honours Degree Program. She has exhibited nationally and internationally, recently in Seoul, Korea, for which she received a British Council Travel Award. She is currently investigating the use of mobile wearable camera technologies to record and transmit live.

Johanna Drucker is the Robertson Professor of Media Studies at the University of Virginia, where she cofounded SpecLab. Her scholarly books include *Sweet Dreams: Contemporary Art and Complicity, The Century of Artists' Books,* and *Alphabetic Labyrinth.*

Robert A. Emmons Jr. is associate director of the Honors College at Rutgers University–Camden, where he teaches film, new media, and comics history. He is a documentary filmmaker and video installation artist. His digital documentaries *Enthusiast, Squeeze, Smalltown USA, YARDSALE!* and *Goodwill: The Flight of Emilio Carranza* focus on features of American popular culture such as comics, the accordion, folk heroes, and small-town life.

Byron Hawk is assistant professor of English at George Mason University and editor of the electronic journal *Enculturation*. He has published articles in the edited volume *The Terministic Screen*, edited by David Blakesley, and the journals *Pedagogy, Technical Communications Quarterly*, and *JAC*. His first book is *A Counter-History of Composition: Toward Methodologies of Complexity*.

Johndan Johnson-Eilola is professor of communication and media at Clarkson University, where he teaches information architecture, new media, technical communication, usability, and mass media. His books include *Datacloud: Toward a New Theory of Online Work, Writing New Media* (with Anne Wysocki, Cindy Selfe, and Geoff Sirc) and *Central Works: Landmark Essays in Technical Communication* (coedited with Stuart Selber).

Richard Kahn is a PhD student in the Social Sciences and Comparative Education division of the Graduate School of Education at the University of California–Los Angeles. He is the ecopedagogy chair of the university's new Paulo Freire Institute.

Douglas Kellner is George Kneller Chair in the Philosophy of Education at the University of California–Los Angeles. He is coauthor or author of many books on social theory, politics, history, and culture, including *Camera Politica: The Politics and Ideology of Contemporary Hollywood Film*; *Postmodern Theory: Critical Interrogations*; *Critical Theory, Marxism, and Modernity*; *Jean Baudrillard: From Marxism to Postmodernism and Beyond*; *Television and the Crisis of Democracy*; and *Grand Theft 2000: Media Spectacle and the Theft of an Election*. With Dan Streible, he coedited *Emile de Antonio: A Reader*.

Karla Saari Kitalong is associate professor of technical communication and Director of Writing Programs at the University of Central Florida in Orlando. Her teaching and research centers on written communication, visual rhetoric, and usability.

Steve Mann is associate professor of electrical and computer engineering at the University of Toronto. In 1991, his inventions and ideas initiated the Massachusetts Institute of Technology Wearable Computing Project. He is regarded as the inventor of the wearable computer and the EyeTap camera, as well as the world's first person to put his personal day-to-day life on the Web as pictures.

Lev Manovich is professor in the visual arts department at the University of California–San Diego, where he teaches new media art and theory. He is the author of *Soft Cinema: Navigating the Database*, *The Language of New Media*, and *Tekstura: Russian Essays on Visual Culture*, as well as more than eighty articles that have been published in twenty-eight countries.

Adrian Miles teaches the theory and practice of hypermedia and interactive video at RMIT University, Melbourne, Australia. He has been a senior new media researcher in InterMedia Lab at the University of Bergen, Norway.

Jason Nolan is assistant professor with the School for Early Childhood Education at Ryerson University in Toronto. He is founding coeditor of the journal *Learning Inquiry* and coeditor of *The International Handbook of Virtual Learning Environments*.

Julian Oliver is a free-software developer, teacher, composer, and media theorist. He has led workshops and master classes in game design, game-engine development, virtual architecture, interface design, augmented reality, and open-source development practices worldwide. In 1998 he established the art-based game-development collective Select Parks.

Ollie Oviedo is associate professor of English at Eastern New Mexico University. His publications include *TnT: Texts and Technology* (coedited with Janice R. Walker) and *The Emerging Cyberculture: Literacy, Paradigm, and Paradox* (coedited with Stephanie Gibson). His essays have been published in several U.S. and international journals. He is editor of *Readerly/Writerly Texts: Essays on Literary, Composition, and Pedagogical Theory*.

Mark Paterson is a lecturer in philosophy and cultural studies at the University of the West of England, Bristol. He is the author of *Consumption and Everyday Life* and has written chapters in *The Book of Touch*, *Emotional Geographies*, and *The Smell Culture Reader*.

Isabel Pedersen is assistant professor of professional communication at Ryerson University in Toronto, and she is affiliated with the Joint Graduate Programme in Communication and Culture at Ryerson University and York University, also in Toronto. She has published in *Semiotica, Social Semiotics*, and *Wascana Review*.

Michael Pennell is assistant professor of writing and rhetoric at the University of Rhode Island. He teaches business communication and writing in electronic environments in the College Writing Program and is currently investigating the literacy education of early Rhode Island mill communities.

Joanna Castner Post is assistant professor of writing at the University of Central Arkansas. She has published articles in *Computers and Composition, CCTE Studies, Technology and English Studies: Innovative Professional Paths*, and *Taking Flight with OWLs: Examining Electronic Writing Center Work*.

Jenny Edbauer Rice is assistant professor of English at the University of Missouri, where she teaches courses on new media writing, public rhetorics, and composition theory. Her work has been published in *Rhetoric Society Quarterly, JAC, Postmodern Culture*, and *Composition Forum*.

David M. Rieder is assistant professor of English at North Carolina State University, where he teaches rhetoric, composition, and technical communication. He is a cofounding editor and current assistant editor of *Enculturation: A Journal for Rhetoric, Writing, and Culture*. His publications include chapters in *Web.Studies: Rewiring Media Studies for the Digital Age* (edited by David Gauntlett) and *The Matrix and Philosophy: Welcome to the Desert of the Real* (edited by William Irwin). He has written for *The Writing Instructor* and *Technical Communication Quarterly*.

Teri Rueb is associate professor in the graduate department of Digital + Media at the Rhode Island School of Design. She was an early pioneer in using global positioning systems (GPS) to create interactive art installations and sound walks in urban and remote landscapes.

James J. Sosnoski is the author of *Token Professionals and Master Critics* and *Modern Skeletons in Postmodern Closets*, as well as various essays on instructional technology, computer-assisted pedagogy, and online collaboration. He has coedited several issues of *Works and Days*.

Lance Strate is professor of communication and media studies at Fordham University. He is president of the Media Ecology Association and author of *Echoes and Reflections: On Media Ecology as a Field of Study*. He has coedited several anthologies, including *Communication and Cyberspace: Social Interaction in an Electronic Environment* and *The Legacy of McLuhan*.

Jason Swarts is associate professor of technical and professional communication at North Carolina State University, where he teaches courses in usability testing, information design, and technical writing. His recent writing on personal digital assistants (PDAs) and cooperative work has been published in *IEEE Transactions on Professional Communication* and *Written Communication*.

Barry Wellman, FRSC, is S. D. Clark Professor of Sociology at the University of Toronto, where he directs the NetLab. Wellman founded the International Network for Social Network Analysis in 1976. In addition to more than two hundred coauthored articles, he has coedited *The Internet in Everyday Life*, *Networks in the Global Village*, and *Social Structures: A Network Approach*.

Sean D. Williams is associate professor of English and associate dean of the Graduate School at Clemson University in South Carolina, where he conducts research on information design, technology, visual communication, and most recently 3-D virtual worlds. His research has appeared in several books and journals including *JAC*, *Computers and Composition*, *Technical Communication Quarterly*, *Business Communication Quarterly*, and *Technical Communication*.

Jeremy Yuille is an interaction designer working with interdisciplinary teams in the fields of communications, education, networked art, and performance. He coordinates digital media courses for the RMIT University, Melbourne, Communication Design program and manages the Multiuser Environments program for the Australasian Cooperative Research Centre for Interaction Design (ACID).